Preconditioning and the Conjugate Gradient Method in the Context of Solving PDEs

Spotlights

Josef Málek and Zdeněk Strakoš, *Preconditioning and the Conjugate Gradient Method in the Context of Solving PDEs*

Preconditioning and the Conjugate Gradient Method in the Context of Solving PDEs

Josef Málek
Zdeněk Strakoš

Charles University in Prague
Prague, Czech Republic

Society for Industrial and Applied Mathematics
Philadelphia

This work is supported by the ERC-CZ project LL1202 financed by the Ministry of Education, Youth, and Sports of the Czech Republic.

Publisher	David Marshall
Acquisitions Editor	Elizabeth Greenspan
Developmental Editor	Gina Rinelli
Managing Editor	Kelly Thomas
Production Editor	Kelly Thomas
Copy Editor	Kelly Thomas
Production Manager	Donna Witzleben
Production Coordinator	Cally Shrader
Compositor	Scott Collins
Graphic Designer	Lois Sellers

Library of Congress Cataloging-in-Publication Data

Málek, Josef.
  Preconditioning and the conjugate gradient method in the context of solving PDEs / Josef Málek, Zdeněk Strakoš, Charles University in Prague, Prague, Czech Republic.
    pages cm. -- (SIAM spotlights ; 01)
  Includes bibliographical references and index.
  ISBN 978-1-611973-83-9
  1. Boundary value problems–Numerical solutions. I. Strakoš, Zdeněk. II. Title.
  QA379.M34 2015
  515'.3533--dc23
                         2014035238

 is a registered trademark.

Contents

Preface

Our times can be characterized by, among many other attributes, the seemingly increasing speed of everything. Within science, it has led to the publication explosion, which reflects the fact that, using the words of Cornelius Lanczos in his essay *The Inspired Guess in The History of Physics*, published in 1964 [119], "science is big business." Under the present circumstances it therefore has become increasingly difficult to stop and invest time and energy into collaboration across disciplines.

The authors of this text work in different fields, namely, in mathematical modeling and PDE analysis and in computational mathematics with emphasis on matrix computations. Given the overspecialization of mathematics (and science in general), the probability that they meet at a conference or workshop and would bounce ideas between themselves is generally very low. Fortunately for them, this did happen a couple of years ago, and the present text is one of the consequences. They found working on it useful and enjoyable, with the goal of finding a common language and developing a mutual understanding for each other's work. In this way they have reacted to the feeling which they had during their first professional discussion and which is so clearly expressed in the beautiful, poetic words of the *one true geometrician in the empire* in *The Wisdom of the Sands* by Antoine de Saint-Exupéry [173, Chapter 115]:

> *I would not bid you pore upon a heap of stones, and turn them over and over,*
> *in the vain hope of learning from them the secret of meditation. For on the level*
> *of the stones there is no question of meditation; for that, the temple must have*
> *come into being. But, once it is built, a new emotion sways my heart, and when*
> *I go away, I ponder on the relations between the stones. ...*
>
> *I must begin by feeling love; and I must first observe a wholeness. After that*
> *I may proceed to study the components and their groupings. But I shall not trou-*
> *ble to investigate these raw materials unless they are dominated by something*
> *on which my heart is set. Thus I began by observing the triangle as a whole; then*
> *I sought to learn in it the functions of its component lines. ...*
>
> *So, to begin with, I practise contemplation. After that, if I am able, I anal-*
> *yse and explain. ...*
>
> *Little matter the actual things that are linked together; it is the links that*
> *I must begin by apprehending and interpreting.*

The authors present the result of their effort to others in the hope that at least some part of it could be useful in strengthening common understanding among communities dealing with mathematics of problems that can be encountered on the way from the formulation of the mathematical model and its analysis towards numerical computation of the discretized systems of algebraic equations.

The case study which is used throughout the text is built around the unifying principle of minimizing the energy of a given infinite-dimensional system as well as the associated discretized finite-dimensional system. That naturally links a class of linear elliptic partial differential boundary value problems described later with the well-known conjugate gradient method, which can be derived using an appropriate Hilbert space setting and then discretized in order to obtain its algebraic formulation. Moreover, the conjugate gradient method can be thought of as a model reduction procedure producing a sequence of discrete models of increasing size which match moments of the original infinite-dimensional model. A more detailed description with a chapter-by-chapter overview of this little book will be presented in the introduction with the help of some basic notation given there.

Our approach can perhaps be characterized by the following quotes. Two of them are taken, as the quote above, from the work of Cornelius Lanczos, who personified in a brilliant way the unity of analytic and computational views (and, in addition, an outstanding unity of knowledge of mathematics and physics as well as an extraordinary depth of philosophical thought). The first one points out the need for mutual understanding, and it is taken from the preface of the monograph *Linear Differential Operators*, published in 1961 [120]:

> *To get an explicit solution of a given boundary value problem is in this age of large electronic computers no longer a basic question. The problem can be coded for the machine and the numerical answer obtained. But of what value is the numerical answer if the scientist does not understand the peculiar analytical properties and idiosyncrasies of the given operator?*

The second quote resonates with the authors' personal feeling that it is good to stop for a while and contemplate in order to understand what has been achieved or, perhaps, to give a second thought to some commonly accepted and widely communicated views. It is taken from the essay [119] of Lanczos mentioned above:

> *Once the great mathematician Gauss was engaged in a particularly important investigation, but seemed to make little headway. His colleagues inquired when the publication was to appear. Gauss gave them an apparently paradoxical and yet perfectly correct answer: "I have all the results but I don't know yet how I am going to get them."*

Finally, we have found very relevant to point out the following quote from the seminal paper of John von Neumann and Herman H. Goldstine [191] that is often considered as the beginning of the field of numerical analysis:

> *When a problem in pure or in applied mathematics is "solved" by numerical computation, errors, that is, deviations of the numerical "solution" obtained from the true, rigorous one, are unavoidable. Such a "solution" is therefore meaningless, unless there is an estimate of the total error in the above sense.*
>
> *Such estimates have to be obtained by a combination of several different methods, because the errors that are involved are aggregates of several different kinds of contributory, primary errors. These primary errors are so different from each other in their origin and character, that the methods by which they have to be estimated must differ widely from each other. A discussion of the subject may, therefore, advantageously begin with an analysis of the main kinds of primary errors, or rather of the sources from which they spring.*
>
> *This analysis of the sources of errors should be objective and strict inasmuch as completeness is concerned, ...*

The investigation of *all sources of errors* should be indeed an inherent part of any numerical computation. We illustrate this point within the last chapters of the text.

We gratefully acknowledge the support of the ERC-CZ project LL1202 *Implicitly constituted material models: from theory through model reduction to efficient numerical methods*, financed by the Ministry of Education, Youth and Sports of the Czech Republic. We are indebted to many friends for their feedback and the invaluable suggestions for further readings unknown to us, in particular to Jan Blechta, Miroslav Bulíček, Roland Herzog, Marta Jarošová, Pavel Jiránek, Axel Klawonn, David Silvester, Gerhard Starke, and Mark Steinhauer. Their input has significantly contributed to our effort in writing this text. We wish to express our special thanks to Jörg Liesen, Volker Mehrmann, Ivana Pultarová, and Endré Süli for their encouragement, for many insightful comments on the text, and for pointing out valuable references. We also appreciate many insightful comments of the referees and of the members of the Editorial Board of the SIAM Spotlight series. Jörg Liesen pointed out that our text might be appropriate for this series. We are grateful to him for this excellent idea. Elizabeth Greenspan from SIAM was extremely helpful, kind, and encouraging in making this idea to happen.

The text is as self-contained as possible, but it is inevitably restrictive. We would be grateful for any feedback concerning its contents and presentation as well as for possible comments which can arise from using it as teaching material.

Josef Málek and Zdeněk Strakoš
Prague, July 2014

Chapter 1

Introduction

Numerical solution of partial differential equations (PDEs) is a large, multidisciplinary area of research. Theoretical mathematical disciplines, such as analysis of initial and boundary value PDE problems, functional analysis, and calculus of variations, form the basis for investigation of qualitative questions on existence, uniqueness, and asymptotic behavior of solutions. Discretization of infinite-dimensional problems requires various techniques ranging from function spaces, approximation theory, and quadrature to mesh generation and evaluation of approximation errors. Discretized finite-dimensional problems are solved using tools from matrix computations, which also requires efficient and numerically stable implementations on highly parallel computer architectures. The goal of this book is to argue that in solving challenging problems, all subtasks should be considered as being inseparable parts of the whole solution process.

Numerical investigation of physical phenomena expressed in the form of mathematical models using the language of PDEs is often associated with algebraic iterative computations. The state-of-the-art analysis of algebraic iterative methods can be characterized by a plethora of approaches based on various views. One can find commonly known results that are mathematically formally correct but the application of which is questionable and the associated conclusions are sometimes plainly wrong. This happens when formal statements are based on unrealistic general assumptions which in practice cannot be satisfied. Assuming exact numerical solution of algebraic problems (including, for example, eigenproblems) or preserving global (bi-)orthogonality among the vectors computed using short recurrences may serve as examples. One can also see abstractions which simplify the studied problem and allow one to develop an elegant mathematical theory. The simplifications can at the same time eliminate from consideration fundamental issues which are, for the given problem, substantial. This can be observed, e.g., in linear asymptotic (worst case) operator analysis of Krylov subspace methods. Such analysis does not take into account particular external forces and boundary conditions which are essential in the underlying mathematical formulation. Approximation of infinite-dimensional operators may not be an appropriate tool for investigating approximation of the *solution of the infinite-dimensional equations* involving these operators.

Analysis of computational problems should aim at interconnecting traditionally almost separated parts, ranging from the (engineering, physical, chemical, or biological) set-up of the problem to the computer solutions of the algebraic problem. The matter was instructively expressed by Hackbusch in his foreword to the special issue of *Computing* published in 1995 which was dedicated to adaptivity:

Adaptivity is a further development. Instead of solving problems with a discretization of very high dimension, it is more reasonable to obtain the same solution quality by a lower dimensional but adapted discretization. Adaptivity has created a new paradigm in mathematical computation. In traditional numerical mathematics, the fields "discretization" (e.g., finite element method), its "numerical analysis" (e.g., error estimates), and "solution algorithms" (e.g., solvers for linear systems) are well separated. Adaptive techniques, however, require a combination of all three. For example, the error estimation has become a part of the algorithm. The concrete discretization is now an outgrowth of the algorithm.

Since then, discretization, its numerical analysis, and analysis of algebraic solvers has indeed become part of a concentrated single effort aimed at solving difficult problems. Many ideas have been developed into a genuine mathematical problem-solving computational technology where the individual steps are problem motivated and their deep interconnection is considered in order to achieve the best possible computational efficiency. Multilevel methods combining coarse space components (the fast global communication over the solution domain) with fine local refinements, as well as their analogues in sophisticated algebraic multigrid schemes, domain decomposition techniques with coarse space components, various discretization space enrichments, distributing the global information at individual iteration steps, hierarchical base discretizations, and discretizations using wavelet-based nested iterations give highly developed examples.

Using some underlying iteration scheme with its powerful acceleration, for historical reasons (and most regrettably) called *preconditioning*, is the common feature of existing approaches. This text does not attempt to offer an overview and it does not focus on construction, analysis, and computational performance of specific techniques. Instead it recalls several principal ideas present in various forms in existing efficient approaches, with the aim of highlighting their general validity.

In particular, we will use for this purpose selected topics within two fields: PDEs (and their analysis via the methods of linear functional analysis and calculus of variations) and iterative matrix computations (here represented by the theory and applications of the conjugate gradient (CG) method). Naturally, we will concentrate on the concept of preconditioning in both fields linked in numerical solution of PDEs via finite-dimensional discretizations of infinite-dimensional problems. In order to ensure a clear and accessible description, we focus in each field on subclasses of problems with properties that make them appropriate for analysis.

In the field of PDEs, we consider a subclass of linear elliptic second order boundary-value problems (BVPs) that can be directly analyzed via the Lax–Milgram lemma, which is a powerful yet simple generalization of the Riesz representation theorem, a basic tool of linear functional analysis. We remark that the Lax–Milgram lemma concerns well-posedness of the problem: Given $b \in V^{\#}$ and a linear bounded operator $\mathscr{A} : V \to V^{\#}$, find $u \in V$ so that the functional equation

$$\boxed{\mathscr{A}\, u = b} \tag{1.1}$$

is satisfied. Here V is a (real infinite-dimensional) Hilbert space, and $V^{\#}$ is its dual consisting of all linear continuous (bounded) functionals from V to $\mathbb{R}$. For the most part we specialize even further and consider only linear second order elliptic PDEs that generate self-adjoint operators, i.e., where $\mathscr{A}$ satisfies the symmetry relation $\langle \mathscr{A}\, u, v \rangle = \langle \mathscr{A}\, v, u \rangle$ for all $u, v \in V$, with $\langle \cdot, \cdot \rangle$ denoting the associated duality pairing, $\langle \cdot, \cdot \rangle : V^{\#} \times V \to \mathbb{R}$.

In the field of matrix computations (numerical linear algebra) we investigate the following problem: Given $\mathbf{b} \in \mathbb{R}^N$ and $\mathbf{A} = (A_{ij}) \in \mathbb{R}^{N \times N}$, find $\mathbf{x} \in \mathbb{R}^N$ so that the system of linear algebraic equations

$$\boxed{\mathbf{Ax} = \mathbf{b}} \qquad (1.2)$$

is satisfied. Consistent with the PDE setting, we focus mainly on matrices that are symmetric and positive definite. As mentioned above, we consider iterative solution of the algebraic problem (1.2) via CG as a representant of the Krylov subspace methods.

As we deal with numerical computations, we have to deal with *approximations* to the solution u of the functional problem (1.1) and to the solution $\mathbf{x}$ of the algebraic problem (1.2). Measuring accuracy of the computed approximations, i.e., the description of computational errors, represents a fundamental issue which needs to be addressed in order to balance discretization and iterative computation and to achieve the maximal possible computational efficiency.

The setting (1.1) provides one of many possible definitions of solution to BVPs described by linear elliptic PDEs. Concepts such as classical solution or distributional solution may serve as two other examples. There are, however, reasons that make (1.1) preferable to others.

In particular, the formulation of (1.1) as

$$\text{to find } u \in V : \qquad \langle \mathscr{A} u, v \rangle = \langle b, v \rangle \qquad \text{for all } v \in V \qquad (1.3)$$

is the basis for the abstract Galerkin method that is used for determining a finite-dimensional approximation to u, and that includes as particular cases the widely used finite element methods, spectral methods, etc. Moreover, there is a natural fundamental principle that makes the formulations (1.1), (1.3) different from other concepts. Indeed, defining on V the functional

$$J(v) := \frac{1}{2} \langle \mathscr{A} v, v \rangle - \langle b, v \rangle, \qquad v \in V, \qquad (1.4)$$

it is not difficult to observe (see Section 3.2 below) that the statement

$$u \in V \text{ solves the functional equation (1.1)}$$

is in the considered setting equivalent to

$$u \in V \text{ minimizes the functional } J \text{ over } V.$$

In general, when introducing the concept of solution, it is essential to take into account any such additional property that can be identified within the problem. As an example, there are many problems in mathematical theory of classical linearized elasticity that can be placed within the setting (1.1) and where the corresponding J represents the functional of potential energy. Then finding the solution to (1.1), (1.3) is equivalent to determining $u \in V$ that minimizes the corresponding *energy functional*; see, e.g., the textbook [75, Section 2.2.1] and the monograph [142, Chapters 5 and 7], which builds upon earlier works by Courant, Friedrichs, von Mises, Prager, Synge and, in particular, Washizu [193], etc. This terminology, originating from structural mechanics, is often generalized, and J of the form (1.4) is called the *energy functional* associated with (1.1), (1.3) even if the problem is stated outside the given physical context. In this study, we exploit the equivalence of the formulations using (1.3) and (1.4) stated above.

The Galerkin method considers a finite-dimensional subspace $V_h \subset V$ and produces, through the formulation

$$\text{to find } u_h \in V_h : \qquad \langle \mathscr{A} u_h, v \rangle = \langle b, v \rangle \qquad \text{for all } v \in V_h$$

or, equivalently,

to find $u_h \in V_h$ minimizing the functional J over V_h,

the finite-dimensional discretization of the original infinite-dimensional problem (1.1), (1.3). Given a basis of the finite-dimensional subspace V_h, this leads to the linear algebraic system (1.2) for the coefficients of the expansion of u_h in this basis. If the algebraic CG is applied to solving the resulting system, then its nth iteration minimizes the discrete analogue of the energy functional (1.4), i.e., the energy norm of the algebraic error, over the n-dimensional algebraic Krylov subspace defined by the matrix $\mathbf{A}$ and the right-hand side $\mathbf{b}$ (here we assume, with no loss of generality, zero initial approximation). The CG method therefore is in the given sense optimal and conforms to the Galerkin discretization of (1.1), (1.3). In order to accelerate convergence, the algebraic problem is often modified which results in the algebraically preconditioned CG method.

One can look at the whole solution process described above, at least in theory, in a different way. Considering the Riesz map τ from $V^{\#}$ to V determined by the choice of some appropriate inner product in V associated with the problem (1.1), (1.3) to be solved, one can form the powers of the composed operator $\tau \mathscr{A}$ with respect to the right-hand-side τb and build up the Krylov subspaces within the infinite-dimensional Hilbert space V. Then the infinite-dimensional Hilbert space formulation of CG gives at the step n immediately the n-dimensional Galerkin discretization of (1.1), (1.3), with the discretization space equal to the nth Krylov subspace generated by $\tau \mathscr{A}$ and τb. The nth CG iteration in the infinite-dimensional Hilbert space then represents the *model reduction* of the original infinite-dimensional problem (1.1), (1.3) to the n-dimensional algebraic problem matching the first $2n$ *moments* of the associated infinite-dimensional distribution function. This makes CG theoretically attractive.

If the infinite-dimensional CG is restricted to some finite-dimensional subspace $V_h \subset V$, then for the fixed basis of V_h the discretization of the infinite-dimensional CG, operating with functions in V_h and functionals in $V_h^{\#}$, directly gives the *preconditioned* CG applied to the algebraic system (1.2) and operating with the coordinates in the bases of V_h and $V_h^{\#}$. The discretization is obtained in a straightforward way without forming, as an intermediate step, the algebraic system (1.2) and the algebraic CG and without preconditioning via the transformation of (1.2). The preconditioning is instead determined by the choice of the basis of V_h and by the choice of the inner product in V_h.

Instead of considering the stages in solving real-world problems ordered linearly as

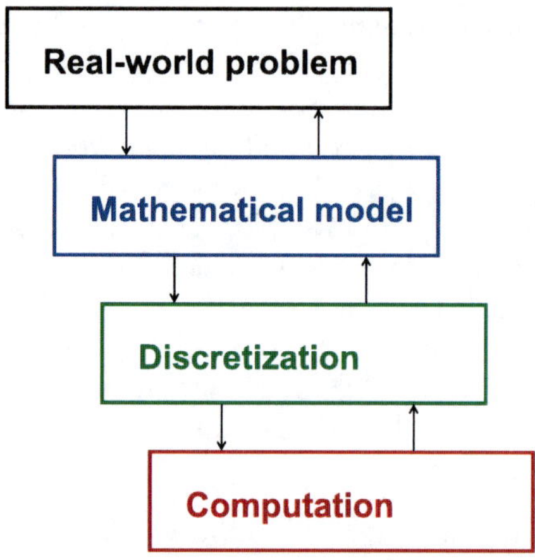

where the arrows indicate the interactions, we therefore argue that, for the case of linear second order elliptic PDEs that generate self-adjoint operators which are bounded and coercive, and for associated iterative computations using preconditioned CG, the following diagram is more relevant:

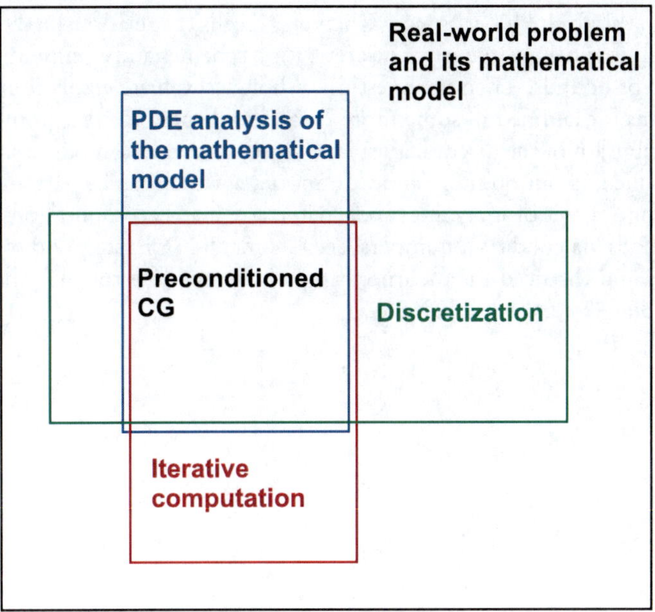

This diagram shows the main idea of the presented text. The preconditioned CG representing iterative computation is in a close and direct interaction with the PDE analysis of the given problem to be solved (through operator preconditioning and model reduction) as well as with the discretization (through the choice of the discretization basis, forming the algebraic system, and, in particular, through the link between the discretization basis and preconditioning).

Although our study is motivated by challenging applications, it does not directly deal with them. Instead it restricts itself to a class of BVP which allows for creation of a unified mathematical structure which forms a basis for the whole text. Nevertheless, we believe that the ideas presented can inspire further developments which will prove useful in important applications.

The structure of the book is as follows. We first consider a class of BVPs associated with second order elliptic linear operators. Considering the link between classical and weak formulations, we rewrite the latter in the form (1.1). We also introduce various equivalent norms on the Sobolev space $H^1(\Omega)$ consisting of functions from the Lebesgue space $L^2(\Omega)$ whose weak partial derivatives of the first order belong to $L^2(\Omega)$. Chapter 3 then recalls several basic results connected with the functional equation (1.1) and its well-posedness: the Riesz representation theorem, a characterization in terms of bilinear forms, and the Lax–Milgram lemma. We also highlight the role of symmetry of bilinear forms and its consequences for the application of methods of the calculus of variations. In Chapter 4 we specify what preconditioning is for the functional equation (1.1). In particular, we recall the concept of operator preconditioning defined by the choice of an appropriate inner product with the associated Riesz map. Chapter 5 presents CG in a general Hilbert space, with the analytic view linking the Vorobyev moment problem, CG, and the Gauss–Christoffel quadrature. Chapter 6 identifies CG in the finite-dimensional

Hilbert space setting with the standard matrix formulation of the preconditioned conjugate gradients (PCG). After introducing the notation for the Galerkin discretization in Chapter 7, the subsequent Chapter 8 provides a general description of the fact that any algebraic preconditioning of the matrix problem can be interpreted as transformation of the discretization basis accompanied with the associated transformation of the inner product in the given Hilbert space. Chapter 9 recalls the relationship between the consistency, stability, and convergence of discretization schemes and comments on using conditioning in error bounds. Then the question of how to evaluate errors in numerical PDE computations is examined in some more detail in Chapter 10. It is pointed out that the spatial distribution of the discretization and algebraic errors can be very different. This fact has remained, in our opinion, almost unnoticed, which can be partially attributed to the prevailing practice of analyzing modern iterative matrix computations using Krylov subspace methods via condition numbers; see Chapter 11. Chapter 12 discusses inexact computations and the need for incorporating all sources of error into the a posteriori analysis. Chapter 13 concludes the book.

Chapter 2

Linear Elliptic Partial Differential Equations

This chapter is written with the objective of making its content (and therefore the rest of the text) accessible to those not working in the field of PDEs. The aim is to describe the family of problems that can be covered by the functional equation (1.1) and, on the other hand, to mention the kind of problems to which the equation (1.1) is not directly applicable. We also focus on various equivalent norms that can be introduced on (the subspaces of) the Sobolev space $H^1(\Omega) := W^{1,2}(\Omega)$ described below and point out some differences between qualitative analysis of PDEs and quantitative numerical approximations of their solutions.

The results presented in this chapter can be found in many books; we refer in particular to [34, 62, 51, 72, 141, 194].

2.1 ▪ Function spaces

First we briefly introduce the function spaces needed in the text below. For an open set $\Omega \subset \mathbb{R}^d$ and $\ell \in \mathbb{N}$, $C^\ell(\Omega)$ denotes the space of all functions continuous in Ω for which any derivative up to order ℓ is continuous in Ω as well; i.e., $D^\alpha u \in C(\Omega)$ for any multiindex $\alpha = (\alpha_1, \ldots, \alpha_d)$, $\alpha_i \in \mathbb{N} \cup \{0\}$, $|\alpha| := \sum_{i=1}^d \alpha_i = \ell$. The space $C^\ell(\overline{\Omega})$ consists of functions u belonging to $C^\ell(\Omega)$ such that for any multiindex α with $|\alpha| \leq \ell$ the function $D^\alpha u$ admits a continuous extension to $\overline{\Omega}$; $C^\infty(\Omega) = \bigcap_{\ell=1}^\infty C^\ell(\Omega)$ and similarly for $C^\infty(\overline{\Omega})$. Finally, $\mathscr{D}(\Omega)$, sometimes also denoted $C_c^\infty(\Omega)$, contains all functions from $C^\infty(\Omega)$ having a compact support in Ω.

Throughout the text vectors with components corresponding to the individual dimensions in $\mathbb{R}^d$ will always be *row vectors*. On the contrary, algebraic vectors associated with the discrete algebraic formulations of various problems using matrix representations will always be *column vectors*. This little ambiguity in notation is unavoidable in order to respect the standard notation in the fields of PDEs and of matrix computations.

Let $(Z, \|\cdot\|_Z)$ be a Banach space.[1] Then $Z^\#$ denotes the *dual space* to Z, which is the space consisting of all linear bounded functionals $\phi : Z \to \mathbb{R}$ equipped with the (dual) norm $\|\phi\|_{Z^\#} := \sup_{\|\phi\|_Z=1} |\langle \phi, z \rangle|$, where $\langle \phi, z \rangle$ stands for the value of ϕ at z. Whenever there are more spaces involved and there is danger of confusion we write $\langle \phi, z \rangle_{Z^\#}$ instead of $\langle \phi, z \rangle$.

---

[1]A normed vector (linear) space $(Z, \|\cdot\|_Z)$ is called a *Banach space* if Z is complete with respect to the norm $\|\cdot\|_Z$.

The *Lebesgue spaces* $L^p(\Omega)$ with the norms $\|\cdot\|_{L^p}$ are defined for $1 \le p \le \infty$ in the following manner. For $1 \le p < \infty$,

$$L^p(\Omega) := \left\{ u : \Omega \to \mathbb{R}; \quad u \text{ is measurable and } \|u\|_{L^p}^p := \int_\Omega |u|^p < \infty \right\};$$

for $p = \infty$ (with $|\Upsilon|_d$ denoting the d-dimensional Lebesgue measure of $\Upsilon \subset \Omega$),

$$L^\infty(\Omega) := \{ u : \Omega \to \mathbb{R}; \quad u \text{ is measurable and}$$
$$\|u\|_{L^\infty} := \inf_{\{\Upsilon \subset \Omega; |\Upsilon|_d = 0\}} \sup_{\{x \in \Omega \setminus \Upsilon\}} \{|u(x)|\} < \infty\}.$$

The *Sobolev spaces* $W^{1,p}(\Omega)$ with the norm $\|\cdot\|_{W^{1,p}}$, where $1 \le p \le \infty$, can be introduced in several ways. Here we proceed via the concept of *weak derivatives*. The Sobolev space $W^{1,p}(\Omega)$ consists of those functions $u \in L^p(\Omega)$ for which all first distributional derivatives belong to (are regular distributions belonging to) $L^p(\Omega)$; i.e., for all $i = 1, \dots, d$ there are $g_i \in L^p(\Omega)$ such that

$$\int_\Omega u \frac{\partial \varphi}{\partial x_i} = -\int_\Omega g_i \varphi \qquad \text{for all } \varphi \in \mathscr{D}(\Omega).$$

It is usual to write $\frac{\partial u}{\partial x_i}$ instead of g_i and ∇u instead of $(g_1, \dots, g_d)$. Then

$$W^{1,p}(\Omega) := \{ u \in L^p(\Omega); \nabla u \in L^p(\Omega)^d \}$$

and (assuming $1 \le p < \infty$)

$$\|u\|_{W^{1,p}} := \left(\|u\|_{L^p}^p + \|\nabla u\|_{L^p}^p \right)^{\frac{1}{p}} = \left(\|u\|_{L^p}^p + \left\| \left(\sum_{i=1}^d \left(\frac{\partial u}{\partial x_i} \right)^2 \right)^{\frac{1}{2}} \right\|_{L^p}^p \right)^{\frac{1}{p}},$$

$$\|u\|_{W^{1,\infty}} := \|u\|_{L^\infty} + \|\nabla u\|_{L^\infty} = \|u\|_{L^\infty} + \inf_{\{\Upsilon \subset \Omega; |\Upsilon|_d = 0\}} \sup_{\{x \in \Omega \setminus \Upsilon\}} \left(\sum_{i=1}^d \left(\frac{\partial u(x)}{\partial x_i} \right)^2 \right)^{\frac{1}{2}}.$$

Sometimes, for $1 \le p < \infty$, the expressions

$$\left(\|u\|_{L^p}^p + \sum_{i=1}^d \left\| \frac{\partial u}{\partial x_i} \right\|_{L^p}^p \right)^{\frac{1}{p}} \qquad \text{or} \qquad \|u\|_{L^p} + \sum_{i=1}^d \left\| \frac{\partial u}{\partial x_i} \right\|_{L^p}$$

are used for the norms on $W^{1,p}(\Omega)$. All these norms are (topologically) equivalent. Note that the norms $\|\cdot\|_X$ and $\|\!|\cdot|\!\|_X$ are (topologically) equivalent on a vector space X if there are $C_1 > 0$ and $C_2 > 0$ such that

$$C_1 \|u\|_X \le \|\!|u|\!\|_X \le C_2 \|u\|_X \qquad \text{for all } u \in X; \tag{2.1}$$

i.e., the norms $\|\cdot\|_X$ and $\|\!|\cdot|\!\|_X$ define in X the same system of open sets. In PDE literature, (2.1) is identified with the equivalence of norms; i.e., the term "equivalence of norms" is used in the meaning of "topological equivalence of norms."

While for the given p the above three norms are equivalent on $W^{1,p}(\Omega)$ with the absolute constants C_1 and C_2 expressed only in terms of p and d, there are many other useful norms equivalent to $\|\cdot\|_{W^{1,p}}$, for which the constants C_1, C_2 depend on the data of the investigated PDE problem, and in many situations they are small and/or large, respectively. This fact has severe consequences in numerical analysis and in particular in the interpretation of its results. We provide more examples of norms equivalent to $\|\cdot\|_{W^{1,p}}$ (with $p = 2$) in Section 2.3 below. The following remark raises an important point.

Remark 1. Topological equivalence of norms represents a standard tool in analysis as well as in numerical analysis of PDEs. Since our text links these fields with algebraic iterative computations needed for finding approximate solutions to the discretized problems, it is important to stress that in *quantitative* considerations on the accuracy of approximations to the solution and on the costs of the whole solution process, the topological equivalence of norms *is not applicable*. Different norms used for measuring the errors can mean significantly different computational cost. Moreover, different parts of the total error should be measured in compatible norms or, more generally, in compatible ways. As outlined in Chapters 9–12 later, this is fundamental for balancing different sources of error within the whole solution process.

In the next section we shall relate the above-defined Sobolev spaces $W^{1,p}(\Omega)$ to the spaces defined via the closure of $C^1(\overline{\Omega})$-functions with respect to the $\|\cdot\|_{W^{1,p}}$ norm. Here, we give just one such definition, namely,

$$W_0^{1,p}(\Omega) := \overline{\mathscr{D}(\Omega)}^{W^{1,p}(\Omega)} := \text{ closure of } \mathscr{D}(\Omega) \text{ in } W^{1,p}(\Omega)$$
$$:= \{ u \in L^p(\Omega); \text{ there exists } \{u_n\}_{n=1}^{\infty} \subset \mathscr{D}(\Omega) \text{ such that}$$
$$\|u_n - u\|_{W^{1,p}} \to 0 \text{ as } n \to \infty \}.$$

Since below we need the Sobolev spaces with $p = 2$ only, we further restrict ourselves to this case and we use the notation that can often be found in literature, namely

$$H^1(\Omega) := W^{1,2}(\Omega) \quad \text{and} \quad H_0^1(\Omega) := W_0^{1,2}(\Omega).$$

Together with $L^2(\Omega)$, these spaces are Hilbert spaces with the inner products

$$(u,v)_{L^2} := \int_{\Omega} u\,v, \qquad (u,v)_{H^1} := \int_{\Omega} (u\,v + \nabla u \cdot \nabla v),$$

where, consistent with the definition of $\|\nabla u\|_{L^p}$ above,

$$\nabla u \cdot \nabla v := \sum_{i=1}^{d} \frac{\partial u}{\partial x_i} \frac{\partial v}{\partial x_i}, \qquad \|u\|_{H^1} = (u,u)_{H^1}^{1/2}.$$

A special case of the above-discussed norm-equivalence is the question under what conditions the H^1 *seminorm* given by

$$|u|_{H^1}^2 := \int_{\Omega} \nabla u \cdot \nabla u = \|\nabla u\|_{L^2}^2$$

represents a norm on certain subspaces of $H^1(\Omega)$. Some of the answers, called Friedrichs and Poincaré inequalities, are given in Section 2.3.

2.2 ■ Setting of the problem: Classical and weak formulation

We wish to formulate a BVP associated with a PDE in a bounded and connected open set $\Omega \subset \mathbb{R}^d$ ($d \in \mathbb{N}$). Formulation of boundary conditions frequently requires an outer normal vector, here denoted by v, defined at (almost) all points of the boundary $\partial\Omega$, i.e., $v = (v_1, \ldots, v_d) : \partial\Omega \to \mathbb{R}^d$. This natural requirement then leads to a certain condition on the smoothness of $\partial\Omega$. Here we assume for simplicity that Ω is a set with the Lipschitz boundary,[2] and we write in short that $\partial\Omega$ is Lipschitz.

Assuming that $\partial\Omega$ consists of two mutually disjoint parts Γ_D and Γ_N, we consider the following problem: Given $u_D : \overline{\Omega} \to \mathbb{R}$, $k, h : \Omega \to \mathbb{R}$, $\sigma, g : \Gamma_N \to \mathbb{R}$, and $\mathbb{K} = (K_{ij})_{i,j=1}^d : \Omega \to \mathbb{R}^{d \times d}$, find $u : \overline{\Omega} \to \mathbb{R}$ satisfying

$$
\begin{aligned}
-\frac{\partial}{\partial x_j}\left(K_{ij}\frac{\partial u}{\partial x_i}\right) + ku &= h && \text{in } \Omega, \\
u - u_D &= 0 && \text{on } \Gamma_D, \\
\frac{\partial u}{\partial \eta} + \sigma u &= g && \text{on } \Gamma_N.
\end{aligned}
\tag{2.2}
$$

Here we have started to use the summation convention. It means that we always take a sum over repeated indices from 1 to d, and thus the first equation in (2.2) in fact means

$$
-\sum_{i,j=1}^d \frac{\partial}{\partial x_j}\left(K_{ij}\frac{\partial u}{\partial x_i}\right) + ku = h \quad \text{in } \Omega.
$$

The symbol $\frac{\partial u}{\partial \eta}$ stands for the derivative with respect to the co-normal vector $\eta : \Gamma_N \to \mathbb{R}^d$ having the components $\eta_i := K_{ij} v_j$, $i = 1, \ldots, d$, which means that

$$
\frac{\partial u}{\partial \eta} = \frac{\partial u}{\partial x_i}\eta_i = K_{ij}\frac{\partial u}{\partial x_i}v_j.
$$

In matrix-vector notation used in algebra we have

$$
\eta = v\mathbb{K}^* \quad \text{and} \quad \frac{\partial u}{\partial \eta} = \nabla u \cdot \eta = (\nabla u)\eta^* = (\nabla u)\mathbb{K}v^* = (\nabla u)\mathbb{K}\cdot v,
$$

where z^* means the transposition of the matrix (vector) z. It should be pointed out that in the algebraic description of the discretized PDEs below, the summation convention is *not* used (it does not represent a standard notation in matrix computations).

The first equation in (2.2) is a scalar linear PDE of second order that becomes *uniformly elliptic* provided that $\mathbb{K}$ meets the following condition:

$$
\begin{aligned}
&\text{there is a } \beta > 0 \text{ such that } K_{ij}(x)\xi_j\xi_i \geq \beta|\xi|^2 \\
&\text{for (almost) all } x \in \Omega \text{ and for all } \xi = (\xi_1, \ldots, \xi_d) \in \mathbb{R}^d.
\end{aligned}
\tag{2.3}
$$

Obviously, if $\mathbb{K}$ is the identity matrix and $k = 0$ in Ω, then we obtain the mixed (Newton) BVP for the Laplace equation that reduces to the (generally nonhomogeneous) Dirichlet

---

[2]We say that $\Omega \subset \mathbb{R}^d$ is a set with Lipschitz boundary, and then write $\partial\Omega$ is Lipschitz, if there is $\ell \in \mathbb{N}$ and the numbers $\alpha_1 > 0$ and $\alpha_2 > 0$ such that the boundary is described by ℓ mutually overlapping Lipschitz maps $\varrho_1, \ldots, \varrho_\ell$, such that, for each map $\varrho \in \{\varrho_1, \ldots, \varrho_\ell\}$, upon appropriately reorienting the coordinate axes, the sets $\{(x_1, \ldots, x_d) \in \mathbb{R}^d; \max_{i=1,\ldots,d-1}|x_i| \leq \alpha_1 \text{ and } \varrho(x_1, \ldots, x_{d-1}) < x_d \leq \varrho(x_1, \ldots, x_{d-1}) + \alpha_2\}$ are subsets of Ω and the sets $\{(x_1, \ldots, x_d) \in \mathbb{R}^d; \max_{i=1,\ldots,d-1}|x_i| \leq \alpha_1 \text{ and } \varrho(x_1, \ldots, x_{d-1}) - \alpha_2 < x_d \leq \varrho(x_1, \ldots, x_{d-1})\}$ are contained in $\mathbb{R}^d \setminus \overline{\Omega}$; see [141] for details.

problem if Γ_N is of Hausdorff (surface) measure zero and to the Neumann problem if Γ_D is of Hausdorff measure zero and $\sigma \equiv 0$ on $\partial\Omega$.

Interestingly, it is also possible to rewrite (2.2) in an equivalent mixed form that in many situations reflects better the (physical) origin of the problem and that reads as follows. Find $u : \Omega \to \mathbb{R}$ and $q : \Omega \to \mathbb{R}^d$ so that

$$
\begin{aligned}
-\operatorname{div} q + k u &= h && \text{in } \Omega, \\
q = (\nabla u)\mathbb{K} \quad \text{or} \quad \nabla u &= q\mathbb{K}^{-1} && \text{in } \Omega, \\
u - u_D &= 0 && \text{on } \Gamma_D, \\
q \cdot \nu + \sigma u &= g && \text{on } \Gamma_N.
\end{aligned}
\tag{2.4}
$$

We will henceforth proceed with the more frequently used formulation (2.2). A few comments on the formulation (2.4) will be presented at the end of Chapter 2.

A classical solution to (2.2) (which means the solution that has at least second derivatives continuous in Ω and meets continuously the boundary conditions stated in (2.2)) requires that the given functions $\mathbb{K}, k, u_D, \sigma, g, h$ are sufficiently smooth as well, which is frequently not the case in applications. Even if these functions are smooth, it is well known (see [89]) that a smooth solution does not exist if the (homogeneous) Dirichlet and (homogeneous) Neumann boundary conditions meet at the boundary or if Ω is a set with a concave corner (such as L-shape domains); see [88]. These are some of the reasons a generalization of the concept of solution to (2.2) is needed, and the theory of weak solutions has been developed since 1930. (See Leray [122] as a remarkable example and [35] for a historical survey.) In order to derive a weak formulation of (2.2) that will indicate how one can weaken the assumptions on the data, we set

$$
\mathcal{V} := \{v \in C^1(\overline{\Omega}); v = 0 \text{ on } \Gamma_D\}.
\tag{2.5}
$$

Multiplying the first equation in (2.2) by an arbitrary $v \in \mathcal{V}$, integrating the result over Ω, using the formula

$$
-\frac{\partial}{\partial x_j}\left(K_{ij}\frac{\partial u}{\partial x_i}\right)v = -\frac{\partial}{\partial x_j}\left(vK_{ij}\frac{\partial u}{\partial x_i}\right) + K_{ij}\frac{\partial u}{\partial x_i}\frac{\partial v}{\partial x_j}
$$

and the Gauss theorem, one arrives at

$$
\int_\Omega K_{ij}\frac{\partial u}{\partial x_i}\frac{\partial v}{\partial x_j} - \int_{\partial\Omega} vK_{ij}\frac{\partial u}{\partial x_i}\nu_j + \int_\Omega kuv = \int_\Omega hv.
\tag{2.6}
$$

Incorporating the boundary conditions (v vanishes on Γ_D) we conclude that u should fulfill

$$
\int_\Omega\left(K_{ij}\frac{\partial u}{\partial x_i}\frac{\partial v}{\partial x_j} + kuv\right) + \int_{\Gamma_N} \sigma uv = \int_{\Gamma_N} gv + \int_\Omega hv \quad \text{for all } v \in \mathcal{V}.
\tag{2.7}
$$

For given $\mathbb{K}, k, \sigma, g$ and h, the left-hand side of (2.7) represents a bilinear form in u and v, while the right-hand side generates a linear form in v. We thus set

$$
\begin{aligned}
a(u,v) &:= \int_\Omega\left(K_{ij}\frac{\partial u}{\partial x_i}\frac{\partial v}{\partial x_j} + kuv\right) + \int_{\Gamma_N} \sigma uv, \\
\langle f,v \rangle &:= \int_{\Gamma_N} gv + \int_\Omega hv
\end{aligned}
\tag{2.8}
$$

and write (2.7) in compact form

$$a(u,v) = \langle f, v \rangle \quad \text{for all } v \in \mathcal{V}. \tag{2.9}$$

Next, taking $v = u - u_D$ in (2.7) (or in (2.9), which we use in order to shorten the notation) we obtain

$$a(u,u) = a(u, u_D) + \langle f, u - u_D \rangle. \tag{2.10}$$

Using the ellipticity condition (2.3) we conclude from (2.8) that

$$a(u,u) \geq \beta \|\nabla u\|_{L^2}^2 + \int_\Omega k|u|^2 + \int_{\Gamma_N} \sigma |u|^2. \tag{2.11}$$

This indicates that the appropriate space where we could look for the solution should be a subspace of $H^1(\Omega)$. We thus define the solution space $H^1_\Gamma(\Omega)$ through

$$
\begin{aligned}
H^1_\Gamma(\Omega) := \overline{\mathcal{V}}^{H^1(\Omega)} &:= \text{ closure of } \mathcal{V} \text{ in } H^1(\Omega) \\
&:= \{u \in L^2(\Omega); \text{ there exists } \{u_n\}_{n=1}^\infty \subset \mathcal{V} \text{ such that} \\
&\qquad \|u_n - u\|_{H^1} \to 0 \text{ as } n \to \infty\}.
\end{aligned}
\tag{2.12}
$$

Different definitions of $H^1(\Omega)$ (via the concept of weak derivatives) and $H^1_\Gamma(\Omega)$ (via the density of smooth functions) require a few comments. Since the boundary $\partial\Omega$ is Lipschitz, two important results are available.

First, for sets with a Lipschitz boundary,[3] it is known that $H^1(\Omega)$ coincides with the closure of $C^1(\overline{\Omega})$ w.r.t. $\|\cdot\|_{H^1(\Omega)}$, i.e.,

$$
\begin{aligned}
H^1(\Omega) = \{u \in L^2(\Omega); \text{ there exists } \{u_n\}_{n=1}^\infty \subset C^1(\overline{\Omega}) \text{ such that} \\
\|u_n - u\|_{H^1} \to 0 \text{ as } n \to \infty\}.
\end{aligned}
$$

In fact, even weaker conditions on $\partial\Omega$ suffice for this type of characterization of $H^1(\Omega)$; see, e.g., [194].

Second, if $\partial\Omega$ is Lipschitz, then there is a linear bounded operator $\gamma : H^1(\Omega) \to L^2(\partial\Omega)$, called the *trace operator*, that generalizes the concept of restriction of a $C(\overline{\Omega})$-function to the boundary to functions from $H^1(\Omega)$.

Consequently, for Ω with a Lipschitz boundary we can characterize the spaces $H^1_0(\Omega)$, $H^1_\Gamma(\Omega)$ and $H^1(\Omega)$ as follows:

$$
\begin{aligned}
H^1_0(\Omega) &= \{u \in H^1; \gamma(u) = 0\}, \\
H^1_\Gamma(\Omega) &= \{u \in H^1(\Omega); \gamma(u)|_{\Gamma_D} = 0\}, \\
H^1(\Omega) &= \overline{C^1(\overline{\Omega})}^{H^1(\Omega)} = \text{ closure of } C^1(\overline{\Omega}) \text{ in } H^1(\Omega).
\end{aligned}
\tag{2.13}
$$

We thus have several Hilbert spaces equipped with the norm $\|\cdot\|_{H^1}$; note that they satisfy

$$H^1_0(\Omega) \subset H^1_\Gamma(\Omega) \subset H^1(\Omega).$$

The boundedness of the linear trace operator $\gamma : H^1(\Omega) \to L^2(\partial\Omega)$ implies that

$$\text{there is a } C_{\mathrm{tr}} > 0: \quad \|\gamma(u)\|_{L^2(\partial\Omega)} \leq C_{\mathrm{tr}}\|u\|_{H^1} \text{ for all } u \in H^1(\Omega). \tag{2.14}$$

---

[3] In fact, it suffices to assume that Ω is a set with continuous boundary; i.e., the functions ϱ in the definition given in the preceeding footnote are continuous, and the other requirements hold.

In fact, since (2.14) can be strengthened to (see [114])

$$\text{there is a } \tilde{C}_{\mathrm{tr}} > 0: \quad \|\gamma(u)\|_{L^2(\partial\Omega)} \le \tilde{C}_{\mathrm{tr}}\|u\|_{L^2}^{\frac{1}{2}}\|u\|_{H^1}^{\frac{1}{2}} \text{ for all } u \in H^1(\Omega),$$

and as $H^1(\Omega)$ is compactly embedded into $L^2(\Omega)$ (see, e.g., [62]), we conclude that

$$\gamma \text{ maps } H^1(\Omega) \text{ into } L^2(\partial\Omega) \text{ compactly.} \tag{2.15}$$

It is known (see [125], [114]) that the image of the trace operator γ, i.e., $\gamma(H^1(\Omega))$, does not coincide with $L^2(\partial\Omega)$. More precisely, there are $g \in L^2(\partial\Omega)$ that cannot be viewed as $\gamma(u)$ where $u \in H^1(\Omega)$. A complete characterization of the space $\gamma(H^1(\Omega))$ is given in Lions and Magenes [125]: if $u \in H^1(\Omega)$, then, using the fractional Sobolev spaces,[4] $\gamma(u) \in H^{1/2}(\partial\Omega) := W^{1/2,2}(\partial\Omega)$ and the trace operator considered as the mapping $\gamma : H^1(\Omega) \to H^{1/2}(\partial\Omega)$ is onto (surjective). Consequently, the inequality (2.14) can be strengthened, and we have

$$\text{there is a } D_{\mathrm{tr}} > 0: \quad \|\gamma(u)\|_{H^{1/2}(\partial\Omega)} \le D_{\mathrm{tr}}\|u\|_{H^1} \text{ for all } u \in H^1(\Omega). \tag{2.16}$$

We can now return to the definition of $H^1_\Gamma := H^1_\Gamma(\Omega)$ (see (2.12)) and impose the assumptions on the data so that all terms in (2.9) are well-defined for $u, v \in H^1_\Gamma$. We suppose that

$$
\begin{aligned}
K_{ij}, k &\in L^\infty(\Omega) \qquad \text{(for all } i,j = 1,\dots,d), \\
\sigma &\in L^\infty(\Gamma_N), \\
u_D &\in H^1(\Omega), \\
h &\in (H^1_\Gamma)^{\#} \\
g &\in \left(H^{1/2}(\Gamma_N)\right)^{\#}.
\end{aligned}
\tag{2.17}
$$

Under these assumptions, we observe that $f \in (H^1_\Gamma)^{\#}$ and $a(\cdot,\cdot)$ is bounded in $H^1(\Omega) \times H^1(\Omega)$. Indeed,

$$
\begin{aligned}
\|f\|_{(H^1_\Gamma)^{\#}} &:= \sup_{v \in H^1_\Gamma; \|v\|_{H^1}=1} |\langle f, v \rangle| \\
&\le \sup_{v \in H^1_\Gamma; \|v\|_{H^1}=1} \left\{ |\langle g, \gamma(v)\rangle_{\left(H^{1/2}(\Gamma_N)\right)^{\#}}| + |\langle h, v\rangle| \right\} \\
&\le \sup_{v \in H^1_\Gamma; \|v\|_{H^1}=1} \left\{ \|g\|_{(H^{1/2}(\Gamma_N))^{\#}}\|\gamma(v)\|_{H^{1/2}(\Gamma_N)} + \|h\|_{(H^1_\Gamma)^{\#}}\|v\|_{H^1} \right\} \\
&\le \sup_{v \in H^1_\Gamma; \|v\|_{H^1}=1} \left\{ D_{\mathrm{tr}}\|g\|_{(H^{1/2}(\Gamma_N))^{\#}}\|v\|_{H^1} + \|h\|_{(H^1_\Gamma)^{\#}}\|v\|_{H^1} \right\} \\
&\le D_{\mathrm{tr}}\|g\|_{(H^{1/2}(\Gamma_N))^{\#}} + \|h\|_{(H^1_\Gamma)^{\#}},
\end{aligned}
\tag{2.18}
$$

---

[4] Recall that with $\alpha \in (0,1)$ the spaces of Hölder continuous functions

$$C^{0,\alpha}(\overline{\Omega}) := \left\{ u \in C(\overline{\Omega}); \ \sup_{x,y\in\Omega; x\neq y} \frac{|u(x)-u(y)|}{|x-y|^\alpha} < \infty \right\}$$

form intermediate spaces between $C(\overline{\Omega})$ and $C^1(\overline{\Omega})$. Similarly, the spaces

$$W^{\alpha,p}(\Omega) := \left\{ u \in L^p(\Omega); \ \frac{|u(x)-u(y)|}{|x-y|^{\alpha+(d/p)}} \in L^p(\Omega \times \Omega) \right\}$$

form intermediate spaces between $L^p(\Omega)$ and $W^{1,p}(\Omega)$. We set $H^\alpha(\Omega) := W^{\alpha,2}(\Omega)$. These spaces can be directly introduced on $\partial\Omega$ using the description of $\partial\Omega$ by means of a finite number of Lipschitz maps.

where we used the trace inequality (2.16). Similarly, by Hölder's inequality and the trace inequality (2.14)

$$
\begin{aligned}
|a(u,v)| &\leq \max_{i,j=1,\dots,d} \|K_{ij}\|_{L^\infty} \|\nabla u\|_{L^2} \|\nabla v\|_{L^2} + \|k\|_{L^\infty} \|u\|_{L^2} \|v\|_{L^2} \\
&\quad + \|\sigma\|_{L^\infty(\Gamma_N)} \|\gamma(u)\|_{L^2(\Gamma_N)} \|\gamma(v)\|_{L^2(\Gamma_N)} \\
&\leq 3\max\{ \max_{i,j=1,\dots,d} \|K_{ij}\|_{L^\infty}, \|k\|_{L^\infty}, C_{\mathrm{tr}}^2 \|\sigma\|_{L^\infty(\Gamma_N)} \} \|u\|_{H^1} \|v\|_{H^1}.
\end{aligned}
\tag{2.19}
$$

To conclude, assuming that $\partial\Omega$ is Lipschitz and $\mathbb{K}$, k, u_D, σ, g, and h fulfill the assumptions (2.3) and (2.17), we say that u is a *weak solution* to the BVP (2.2) if

$$
\begin{aligned}
u - u_D &\in H_\Gamma^1, \\
a(u,v) &= \langle f, v \rangle \quad \text{for all } v \in H_\Gamma^1.
\end{aligned}
\tag{2.20}
$$

Setting $\tilde{u} := u - u_D$ and $\langle b, v \rangle := \langle f, v \rangle - a(u_D, v)$, we rewrite (2.20) in the form

$$
\text{find } \tilde{u} \in H_\Gamma^1: \qquad a(\tilde{u}, v) = \langle b, v \rangle \quad \text{for all } v \in H_\Gamma^1.
$$

It is easy to observe that $b \in (H_\Gamma^1)^\#$. Relabeling $\tilde{u}$ as u, we arrive at the commonly known problem: Given $b \in (H_\Gamma^1)^\#$

$$
\text{find } u \in H_\Gamma^1: \qquad a(u, v) = \langle b, v \rangle \quad \text{for all } v \in H_\Gamma^1.
\tag{2.21}
$$

Finally, noticing that for $u \in H_\Gamma^1$ the linear form $a(u, \cdot) \in (H_\Gamma^1)^\#$, we define

$$
\mathcal{A} : H_\Gamma^1 \to (H_\Gamma^1)^\#
$$

as the mapping that assigns to $u \in H_\Gamma^1$ the functional $a(u, \cdot) \in (H_\Gamma^1)^\#$, and rewrite (2.21) in the following form: Given $b \in (H_\Gamma^1)^\#$,

$$
\text{find } u \in H_\Gamma^1: \qquad \langle \mathcal{A} u, v \rangle = \langle b, v \rangle \quad \text{for all } v \in H_\Gamma^1,
\tag{2.22}
$$

which is the functional equation (1.1) we wished to arrive at. Obviously (some details are given in the next chapter), the functional equation (1.1) and its bilinear form representation (2.21) are equivalent.

The differential formulation (2.2) is sometimes called *classical*, and the integral formulation (2.21) is called *weak*. As indicated, the concepts *weak* solution, *weak* derivatives and *weak* formulation reflect the fact that these concepts are meaningful for much *weaker* assumptions on the data, in comparison to the notions of classical solution and classical formulation. Let us mention that it is not difficult to check that if the data are smooth enough and a weak solution $u \in H_\Gamma^1$ of (2.21) is such that it additionally satisfies $u \in C^2(\Omega) \cap C(\overline{\Omega})$, then u is a classical solution and fulfills the classical formulation (2.2).

On the other hand, the word "weak" may suggest certain imperfection of the concepts of weak solution and weak formulation. This viewpoint, however, does not necessarily reflect the nature of the phenomenon modeled. The integral formulation can be physically more meaningful and it appeared (in physics or geometry) earlier than the classical formulation. From this perspective, the integral (weak) formulation is a primary object and the classical formulation is the secondary tool. A direct approach of the calculus of variations serves as an excellent example.[5] Indeed, if $a(u,v) = a(v,u)$ for all u, $v \in H_\Gamma^1$,

---

[5]A similar comment concerns the balance equations in continuum mechanics and thermodynamics; see, e.g., [145, 122, 66, 36]. It should be mentioned that there are physically based PDE problems that do not have weak (or distributional) solutions and one has to look for more general concepts; see, e.g., the interesting paper [155].

which by (2.8) holds if $\mathbb{K}$ is symmetric, i.e., $\mathbb{K} = \mathbb{K}^*$, one defines the functional $J : H_\Gamma^1 \to \mathbb{R}$ through

$$J(v) := \frac{1}{2}a(v,v) - \langle b, v \rangle. \tag{2.23}$$

Then $u \in H_\Gamma^1$ minimizes J over H_Γ^1 if and only if $u \in H_\Gamma^1$ fulfills the integral formulation (2.21); for details see Section 3.2. Thus the integral (weak) formulation (2.21) is a primary object characterizing the existence of the minimizer of J, while the associated classical formulation, called in this context the Euler or Euler–Lagrange equation, is obtained from (2.21) under additional assumptions on the smoothness of the data and of the solution. Let us note that (2.21) represents the Gâteaux derivative of J at $u \in H_\Gamma^1$ evaluated at any $v \in H_\Gamma^1$ (or the directional derivative of J at u and at any direction v) set to zero, and this is the reason why (2.21) is frequently called a *variational formulation*.

The term "variational formulation" associated with quadratic functionals of the form (2.23) will appear in this text in several contexts. In particular, CG as the main representative of Krylov subspace methods for computing numerical approximations to the solution u of (1.1) (which is practically done via computing numerical approximations $\mathbf{x}$ to the solution of the algebraic system (1.2)) can be associated with minimizing (2.23) over Krylov subspaces; see the seminal paper by Hestenes and Stiefel [96], the contribution of Rutishauser in [60] and the infinite-dimensional and finite-dimensional descriptions in Chapters 5 and 6. Therefore CG can be viewed as a *variational method*; for some early comments on this point see [174, Part II, Chapter 5]. As described in Chapters 5 and 6, this link of the PDE formulation (1.1) or (2.21) to the numerical approximation given by CG is of fundamental importance. CG applied to (1.1) solves at the nth iteration step the reduced n-dimensional model determined by the orthogonal projection of the original model represented by (1.1) onto the nth Krylov subspace.

2.3 ▪ Norms equivalent to $\|\cdot\|_{H^1}$: Friedrichs and Poincaré inequalities

The PDE problem (2.2) reformulated as the weak formulation (2.21) can be investigated by various techniques: the Galerkin (energy) method, direct methods of the calculus of variations, representation of the inverse of the differential operator by the associated Green function (see, e.g., [76, Chapters 9–11], [62, Section 2.2.4]), and first order system least squares method (FOSLS; see, e.g., [26]), are just a few examples. The Lax–Milgram lemma, which we shall use in this text (see the next chapter), is one of the popular approaches.

It requires that the bilinear form $a(\cdot,\cdot)$ is *bounded* (which has already been proved; see (2.19)) and H_Γ^1-*elliptic*, which means that[6]

$$\text{there is an } \alpha > 0: \qquad a(u,u) \geq \alpha\|u\|_{H_\Gamma^1}^2 \text{ for all } u \in H_\Gamma^1. \tag{2.24}$$

Recall (see (2.13)) that H_Γ^1 satisfies $H_0^1 \subset H_\Gamma^1 \subset H^1$ with $H_\Gamma^1 = H_0^1$ when $\Gamma_N = \emptyset$ and $H_\Gamma^1 = H^1$ when $\Gamma_D = \emptyset$. Also, it follows from the definition of H_Γ^1 (see (2.12)) that

$$\|\cdot\|_{H_\Gamma^1} = \|\cdot\|_{H^1}.$$

---

[6] Equivalently, one can say that

$$\text{there is an } \alpha > 0: \qquad \inf_{u \in H_\Gamma^1; \|u\|_{H_\Gamma^1} = 1} a(u,u) = \alpha > 0.$$

Specification of the conditions that guarantee the validity of (2.24) is the main motivation for writing this section. Besides its importance in the application of the Lax–Milgram lemma (see Chapter 3), the inequality (2.24) together with (2.19) is, for $a(\cdot,\cdot)$ symmetric, tantamount to the equivalence of the $\|\cdot\|_{H^1}$ norm and the norm $\|\cdot\|_a := \sqrt{a(\cdot,\cdot)}$ in the sense given in (2.1). As we shall see below, we can generate plenty of norms (associated with some inner products) on H^1_Γ that, while being (topologically) equivalent to $\|\cdot\|_{H^1}$, fulfill (2.1) with C_1 very small and/or C_2 very large. This indicates that (topological) equivalence of norms can be completely irrelevant in "quantitative" numerical analysis, and an inappropriate choice of the norm in which the (discretization and algebraic) error is measured can have severe impact on the efficiency of computational methods. This is why, in our opinion, a more detailed discussion concerning the validity of (2.24) is an important issue that deserves further attention.

Since the given function $\mathbb{K}$ fulfills (2.3), and assuming further that[7]

$$k \geq 0 \text{ a.e. in } \Omega \quad \text{and} \quad \sigma \geq 0 \text{ a.e. on } \Gamma_N, \tag{2.25}$$

we know (see (2.11)) that

$$a(u,u) \geq \beta\|\nabla u\|^2_{L^2} + \int_\Omega k|u|^2 + \int_{\Gamma_N} \sigma|u|^2 \geq \beta\|\nabla u\|^2_{L^2}.$$

This does not automatically imply (2.24). Indeed, if $k = 0$, $\sigma = 0$, and $\Gamma_D = \emptyset$, then taking any nonzero constant u we see that $a(u,u) = 0$ while $\|u\|_{H^1_\Gamma} = \|u\|_{L^2(\Omega)} > 0$. However, in the setting described in Section 2.2 and by (2.25), $k = 0$, $\sigma = 0$, and $\Gamma_D = \emptyset$ is the only possibility when (2.24) does not hold. More precisely, the H^1_Γ-ellipticity condition (2.24) holds if at least one of the following conditions is satisfied:

(1) k is positive on a subset of Ω of positive Lebesgue measure $|\cdot|_d$, i.e., there is a $U \subset \Omega$ such that $|U|_d > 0$ and $k > 0$ a.e. in U;

(2) σ is positive on a subset of Γ_N of positive $(d-1)$-dimensional Hausdorff measure $|\cdot|_{d-1}$, i.e., there is a $\Sigma \subset \Gamma_N$ such that $|\Sigma|_{d-1} > 0$ and $\sigma > 0$ a.e. in Σ;

(3) Γ_D is not of measure zero, i.e., $|\Gamma_D|_{d-1} > 0$.

Argumentum ad absurdum of this assertion (see, e.g., [141, Section 2.7]) starts by assuming that at least one of the conditions (1)–(3) is satisfied and the inequality (2.11) and the negation of (2.24) hold. This leads to the existence of a sequence $\{u_n\} \subset H^1_\Gamma$, which can be (without any loss of generality) normalized so that

$$\|u_n\|_{H^1} = 1, \tag{2.26}$$

satisfying

$$\beta\|\nabla u_n\|_{L^2} + \int_\Omega k|u_n|^2 + \int_{\Gamma_N} \sigma|u_n|^2 \leq a(u_n,u_n) \leq \frac{1}{n}. \tag{2.27}$$

---

[7]The abbreviation "a.e. in Z" stands for *almost everywhere in Z*. This means that the considered property holds for all $x \in Z \setminus \Upsilon$, where the Lebesgue (volume) or the Hausdorff (surface) measure of Υ is zero, depending on whether Z is a subset of Ω or $\partial\Omega$, respectively.

As the closed unit ball in the infinite-dimensional separable Hilbert space $H^1(\Omega)$ is weakly compact, it follows from (2.26) that there is a subsequence $\{u_{n_m}\} \subset \{u_n\}$ and $u \in H^1_\Gamma$ so that

$$u_{n_m} \text{ converges to } u \text{ weakly in } H^1(\Omega) \text{ as } m \to \infty$$

$$\text{and by (2.27)} \quad \|\nabla u\|_{L^2} \leq \liminf_{m \to \infty} \|\nabla u_{n_m}\|_{L^2} \leq \limsup_{m \to \infty} \|\nabla u_{n_m}\|_{L^2} \leq 0. \tag{2.28}$$

Referring to the compact embedding of $H^1(\Omega)$ into $L^2(\Omega)$ (see, e.g., [62]), the first line of (2.28) implies that

$$u_{n_m} \text{ converges to } u \text{ strongly in } L^2(\Omega) \text{ as } m \to \infty$$

$$\text{and consequently} \quad \|u\|_{L^2} = \lim_{m \to \infty} \|u_{n_m}\|_{L^2}. \tag{2.29}$$

Since the trace operator γ maps $H^1(\Omega)$ compactly into $L^2(\partial\Omega)$ (see (2.15))

$$u_{n_m} \text{ converges to } u \text{ strongly in } L^2(\partial\Omega) \text{ as } m \to \infty. \tag{2.30}$$

Hence

$$\int_\Omega k|u|^2 + \int_{\Gamma_N} \sigma|u|^2 = \lim_{m \to \infty} \left(\int_\Omega k|u_{n_m}|^2 + \int_{\Gamma_N} \sigma|u_{n_m}|^2 \right) \leq 0. \tag{2.31}$$

By (2.28), $\nabla u = 0$ a.e. in Ω, which implies that u as a function from $H^1(\Omega)$ is *constant*. Also, it follows from (2.28) and (2.29) that

$$\|u\|^2_{L^2} = \|u\|^2_{L^2} + \|\nabla u\|^2_{L^2} = \lim_{m \to \infty} (\|u_{n_m}\|^2_{L^2} + \|\nabla u_{n_m}\|^2_{L^2}) = \lim_{m \to \infty} \|u_{n_m}\|^2_{H^1} = 1. \tag{2.32}$$

While u being constant (almost) everywhere in Ω and Ω is connected, any of the conditions (1)–(3) and (2.31) imply that u is zero (almost) everywhere in Ω, which, however, contradicts (2.32) (as we wished to show).

Three observations are in order.

(i) Inspired by the conditions (1)–(3), we present an illustrative list of norms that are (topologically) equivalent to the $\|\cdot\|_{H^1}$ norm and satisfy (2.1) with arbitrary $C_1 \in (0, 1]$ and $C_2 \geq 1$. Let $\mathbb{L} = (L_{ij})^d_{i,j=1}$ be symmetric positive definite and let us assume for simplicity that L_{ij} is constant in Ω for all $i,j = 1,\ldots,d$. Then

$$\lambda_{\min}|\xi|^2 \leq L_{ij}\xi_i\xi_j \leq \lambda_{\max}|\xi|^2,$$

where $0 < \lambda_{\min} \leq \lambda_{\max}$ denote the smallest and the largest eigenvalue of $\mathbb{L}$. It follows from the discussion above that the norms (we label them by the capital roman numbers)

$$\|u\|^2_I := \int_\Omega L_{ij} \frac{\partial u}{\partial x_i} \frac{\partial u}{\partial x_j} + 10^m \int_{\Omega_0} u^2 \qquad (\Omega_0 \subset \Omega; |\Omega_0|_d > 0),$$

$$\|u\|^2_{II} := \int_\Omega L_{ij} \frac{\partial u}{\partial x_i} \frac{\partial u}{\partial x_j} + 10^m \int_{\Sigma_0} u^2 \qquad (\Sigma_0 \subset \Gamma_N; |\Sigma_0|_{d-1} > 0),$$

$$\|u\|^2_{III} := \int_\Omega L_{ij} \frac{\partial u}{\partial x_i} \frac{\partial u}{\partial x_j} + 10^m \int_\Omega u^2$$

are equivalent to $\|\cdot\|_{H^1}$ on H^1_Γ. Here $\lambda_{\min}$ and $\lambda_{\max}$ as well as the integer m modify the properties of the norms in relation to $\|\cdot\|_{H^1}$. Thus, for example, by an appropriate choice of $\lambda_{\min}$, $\lambda_{\max}$, and m the norms $\|\cdot\|_{H^1}$ and $\|\cdot\|_{III}$ fulfill the norm equivalence condition (2.1) with $C_1 > 0$ very small and/or $C_2 > 1$ very large.

Also, if Γ_D is nonempty, then the expression

$$\|u\|_{IV} := \left(\int_\Omega L_{ij} \frac{\partial u}{\partial x_i} \frac{\partial u}{\partial x_j} \right)^{\frac{1}{2}}$$

is the norm equivalent to $\|\cdot\|_{H^1}$ on H^1_Γ. Thus, in particular, considering $\mathbb{L} = \mathbb{I}$ we conclude that $\|u\|_{H^1}$ and $|u|_{H^1}$ are norms equivalent on H^1_Γ; see also the comments concerning the Friedrichs inequality below.

(ii) If none of the above conditions (1)–(3) hold, which means, that $k = 0$, $\sigma = 0$, and $\Gamma_D = \emptyset$, then we are investigating the Neumann problem:

$$-\frac{\partial}{\partial x_j} \left(K_{ij} \frac{\partial u}{\partial x_i} \right) = h \qquad \text{in } \Omega,$$
$$\frac{\partial u}{\partial \eta} = g \qquad \text{on } \partial\Omega. \tag{2.33}$$

As was mentioned above, (2.24) does not hold in this situation. One possibility for overcoming this problem is to restrict the considered functions to H^1 functions with zero mean value over Ω. Then the space where the solution is sought is given by

$$H^1_{\text{mean}} := \left\{ v \in H^1(\Omega); \int_\Omega v = 0 \right\},$$

and a minor modification of the proof given above implies that the bilinear form $a(u, v)$ is H^1_{mean}-elliptic; see [141, section 1.3]. The Neumann problem (2.33) does not have a solution for arbitrary data g and h. It is necessary that g and h fulfill certain compatibility conditions; otherwise the solution to problem (2.33) does not exist. To recall this, note that (2.33) can be written as

$$-\operatorname{div} q = h, \qquad q = \nabla u \mathbb{K} \qquad \text{in } \Omega,$$
$$q \cdot v = g \qquad\qquad\qquad \text{on } \partial\Omega. \tag{2.34}$$

By the Gauss theorem, $\int_\Omega \operatorname{div} q = \int_{\partial\Omega} q \cdot v$, and we obtain the following compatibility condition:

$$\int_\Omega h + \int_{\partial\Omega} g = 0.$$

(iii) The proof of (2.24) given above is by contradiction. Consequently, a direct proof of (2.24) may be preferable. If $H^1_\Gamma = H^1_0$ (the Dirichlet problem), it is known (see, e.g., [141]) that the following Friedrichs inequality holds:

$$\text{there is a } C_f > 0 : \|u\|_{L^2} \le C_f |u|_{H^1} \text{ i.e. } \|u\|_{L^2} \le C_f \|\nabla u\|_{L^2} \text{ for all } u \in H^1_0. \tag{2.35}$$

With this inequality (2.24) follows easily. Indeed, by (2.11),

$$a(u, u) \ge \frac{\beta}{2} \|\nabla u\|_{L^2}^2 + \frac{\beta}{2} \|\nabla u\|_{L^2}^2 \ge \frac{\beta}{2} \|\nabla u\|_{L^2}^2 + \frac{\beta}{2C_f^2} \|u\|_{L^2}^2$$

$$\ge \alpha \|u\|_{H^1}^2 \qquad \text{with } \alpha := \min\{\beta/2, \beta/(2C_f^2)\}.$$

Similarly, if $H_{\Gamma}^1 = H^1$ (the Neumann or Newton problem, depending on whether $\sigma = 0$ or not), then the following Poincaré inequality holds (see, e.g., [62]):

$$\text{there is a } C_p > 0: \qquad \left\| u - \frac{1}{|\Omega|_d} \int_\Omega u \right\|_{L^2} \leq C_p |u|_{H^1} \qquad \text{for all } u \in H^1. \tag{2.36}$$

The Poincaré inequality implies the equivalence of $\|\cdot\|_{H^1}$ and the $|\cdot|_{H^1} = \|\nabla \cdot\|_{L^2}$ on the space H^1_{mean}.

2.4 ▪ On (u,q) formulation and other possible generalizations

The alternative formulation (2.4) transforms the *scalar second order* PDE problem into *the system of the first order* PDEs. Obviously, the concept of solution to the problem (2.4) involves u as well as the flux q, which might be of particular interest. Despite many important studies on this topic, formulations of the type (2.4) are, in our opinion, of interest for further detailed investigation due to the following reasons.

First, the formulation (2.4) represents the simplest version of the setting where q and ∇u are related implicitly[8] through a given function $r : \mathbb{R}^d \times \mathbb{R}^d \to \mathbb{R}^d$; this means that instead of the first two equations in (2.4) (which are equivalent to the first equation in (2.2)) we have

$$\begin{aligned} -\operatorname{div} q + ku &= h \qquad \text{in } \Omega, \\ r(q, \nabla u) &= 0 \qquad \text{in } \Omega. \end{aligned} \tag{2.37}$$

If q in the second equation cannot be expressed as a function of ∇u, there seems to be no other way than to deal with (2.37) as it stands.

Second, instead of $r(q, \nabla u) = 0$ one can have, in addition to $-\operatorname{div} q + ku = h$ or its time-dependent variants such as

$$\frac{\partial u}{\partial t} - \operatorname{div} q + ku = h \qquad \text{or} \qquad \frac{\partial^2 u}{\partial t^2} - \operatorname{div} q + ku = h,$$

an evolutionary partial differential equation for q. For all these problems, the understanding of the mathematical properties associated with (2.4) is crucial.

Since for practical reasons this text presents ideas that link PDEs through functional analysis to matrix computation within the simplest setting possible, we do not consider the (u,q) formulation further in this study, and we refer the reader to, e.g., [28, 27, 9, 6] for results concerning the treatment of such so-called mixed formulations and to, e.g., [40, 41] for using the first order least squares method to analyze (numerically and computationally) the problem (2.4). It is also worth mentioning that if $\mathbb{K}$ is symmetric and consequently $a(u,v)$ is symmetric, then the problem (2.4) can be treated by the methods of calculus of variations in the *primal* form for u (as discussed above), or in the *dual* form

---

[8]While *implicit constitutive theories* have been used in viscoelasticity and plasticity for a long time, their development in a systematic manner and for more general classes of materials that includes elastic, viscous, viscoelastic, and inelastic materials, as well as a very general implicit functional relationship between the history of the stress and the history of the deformation, with a a thermodynamic footing, are due to the seminal works of Rajagopal [160, 161]. Such an implicit theory extends the standard constitutive theory so that the framework is sufficiently robust to describe complicated nonlinear responses of materials. It also provides an appropriate theoretical basis for the derivation/justification of many models that have been used, mostly successfully, in various areas (engineering, chemistry, food processing) but which have been proposed in an ad hoc manner. It concerns both fluid mechanics (see, e.g., [165, 128, 37, 156]) and solid mechanics (see, e.g., [162, 164, 166, 68, 163]).

for q, or in the *mixed* form for u and q. In this regard we refer to the classical books [57, 185] or to the article [5].

For the same reasons, so as to keep the setting as simple as possible in order to highlight the connections from PDE theory through functional analysis to matrix computations (while limiting the length of the text), we do not pay further attention to various other generalizations. Note, e.g., that not all linear second order elliptic problems are covered by the setting presented in Section 2.2. There are (at least) five essential assumptions that outlined the considered setting, namely

(i) the PDEs considered are scalar;

(ii) $\mathbb{K}$ is uniformly elliptic (see (2.3));

(iii) the additional term $c_j \frac{\partial u}{\partial x_j}$ with a given $c \in L^\infty(\Omega)^d$ is not included in the formulation (2.2);[9]

(iv) k is nonnegative;

(v) σ is nonnegative.

In general, H^1_Γ-ellipticity (2.24) may fail to hold if any of the assumptions (ii)–(v) is not met. It is fair to admit that the omitted cases include important classes of problems.

For example, for the *system* of s linear elliptic PDEs of the form

$$-\frac{\partial}{\partial x_j}\left(K^{kl}_{ij}\frac{\partial u_k}{\partial x_i}\right) = f_l, \quad l = 1, 2, \dots s, \tag{2.38}$$

one replaces the uniform ellipticity condition (2.3) with the condition

$$\text{there is } \tilde{\beta} > 0 \text{ such that } K^{kl}_{ij}(x)\eta_{ik}\eta_{jl} \geq \tilde{\beta}|\eta|^2$$
$$\text{for (almost) all } x \in \Omega \text{ and for all } \eta \in \mathbb{R}^d \times \mathbb{R}^s. \tag{2.39}$$

If this condition is weakened and one requires that

$$K^{kl}_{ij}(x)\eta_{ik}\eta_{jl} \geq 0 \qquad \text{for (almost) all } x \in \Omega \text{ for all } \eta \in \mathbb{R}^d \times \mathbb{R}^s, \tag{2.40}$$

upon adding the term $c^{kl}_j \frac{\partial u_k}{\partial x_j}$ to (2.38), the studied system

$$-\frac{\partial}{\partial x_j}\left(K^{kl}_{ij}\frac{\partial u_k}{\partial x_i}\right) + c^{kl}_j\frac{\partial u_k}{\partial x_j} = f_l, \quad l = 1, 2, \dots s,$$

can include the Stokes problem: Given $f : \Omega \to \mathbb{R}^d$, find $(v, p) : \Omega \to \mathbb{R}^d \times \mathbb{R}$ satisfying

$$-\Delta v + \nabla p = f \text{ in } \Omega, \quad \operatorname{div} v = 0 \text{ in } \Omega, \quad v = 0 \text{ on } \partial\Omega. \tag{2.41}$$

This is a system of $d + 1$ equations for $v = (v_1, \dots, v_d)$ and p, in which the second order operator in the equation $\operatorname{div} v = 0$ is missing (and consequently (2.39) is not met, while

---

[9]Note that $\frac{\partial}{\partial x_j}(K_{ij}\frac{\partial u}{\partial x_i})$ includes $\frac{\partial K_{ij}}{\partial x_j}\frac{\partial u}{\partial x_i}$, which is *formally* of the form $c_j\frac{\partial u}{\partial x_j}$, but the requirement that $c \in L^\infty(\Omega)^d$ requires that $\frac{\partial K_{ij}}{\partial x_j}$ belongs, for each $i = 1, \dots, d$, to $L^\infty(\Omega)$. This is not, however, assumed in the setting given in Section 2.2; see (2.17).

(2.40) holds). One can alternatively proceed similarly as in the case of the Neumann problem discussed above, consider the closed subspace H^1_{div} of $H^1_0(\Omega)^d$ defined through

$$H^1_{\mathrm{div}} := \{u = (u_1, \ldots, u_d) \in H^1_0(\Omega)^d ; \operatorname{div} u = 0\}$$

and observe (in analogy to (2.24)) that the bilinear form

$$(\nabla u, \nabla u)_{L^2} := \frac{\partial u_i}{\partial x_j} \frac{\partial u_i}{\partial x_j}$$

is H^1_{div}-elliptic.

Omitting the (convective) term $c_j \frac{\partial u}{\partial x_j}$ is certainly restrictive from the point of view of many applications. Note, however, that if $\|c\|_\infty$ is small or if c of any size satisfies $\operatorname{div} c = 0$, then the term $c_j \frac{\partial u}{\partial x_j}$ can be incorporated into the setting of Section 2.2.

Finally, within our setting k and σ are supposed to be nonnegative, see (2.25). Consequently, if $\lambda \geq 0$, then the operator $-\Delta + \lambda$ (with, e.g., a homogeneous Dirichlet boundary condition) is admissible. On the other hand, the operator $-\Delta - \lambda$ with $\lambda \geq \lambda_1$, where λ_1 is the smallest eigenvalue of the Dirichlet problem for the Laplace operator, is excluded from the setting considered here. Both operators are called the *Helmholtz operators* in the literature. Most frequently this terminology is associated with $-\Delta - \lambda$. Such problems are usually treated via the Fredholm alternative. Similar comments concern the problem with σ negative (and large enough).

All these and many other extensions are important and should be subsequently investigated within the context presented in this study, which, for the sake of clarity and length limitations, focuses on the simplest (yet broad enough) framework described in Section 2.2.

Chapter 3

Elements of Functional Analysis

We now return to the abstract linear functional analysis setting of the problem (1.1), with the particular formulations of Chapter 2 serving as examples. In order to keep the text self-contained as much as possible, this chapter contains mostly standard textbook material. For general references we suggest, e.g., [62], [76], [30], and the summary in the introductory parts of [103].

3.1 ▪ Riesz representation theorem

Let V be a real Hilbert space with the inner product $(\cdot,\cdot)_V : V \times V \to \mathbb{R}$ and the associated norm $\|\cdot\|_V := \sqrt{(\cdot,\cdot)_V}$. Further let $V^\#$ denote the dual space of bounded (continuous) linear functionals on V with the duality pairing

$$\langle\cdot,\cdot\rangle : V^\# \times V \to \mathbb{R}$$

and the dual norm

$$\|f\|_{V^\#} = \sup_{v \in V; \|v\|_V = 1} |\langle f,v\rangle|.$$

The Riesz representation theorem provides an isometric isomorphism between V and $V^\#$ given through the *Riesz map* $\tau : V^\# \to V$. For each $f \in V^\#$ there exists a unique $\tau f \in V$ such that

$$\langle f,v\rangle = (\tau f,v)_V \quad \text{for all } v \in V. \tag{3.1}$$

We immediately observe that

$$\|\tau f\|_V = \|f\|_{V^\#}. \tag{3.2}$$

Indeed,

$$\|f\|_{V^\#} = \sup_{v \in V; \|v\|_V = 1} |\langle f,v\rangle| = \sup_{v \in V; \|v\|_V = 1} |(\tau f,v)_V| \le \|\tau f\|_V.$$

Taking $v = \tau f / \|\tau f\|_V$ in (3.1) gives the equality.

Example 3.1. Consider $H_0^1 := H_0^1(\Omega)$ (see (2.13)) with the inner product $(\cdot,\cdot)_{H^1}$. The Riesz map

$$\tau : H^{-1} := (H_0^1)^\# \to H_0^1$$

associated with this inner product is given by

$$\langle f, v \rangle = (\tau f, v)_{H^1} = \int_\Omega \{ (\tau f) v + \nabla(\tau f) \cdot \nabla v \} \quad \text{for all } f \in H^{-1} \text{ and for all } v \in H_0^1.$$

Denoting $u = \tau f$, we observe that, for the given f, u fulfills

$$\int_\Omega (uv + \nabla u \cdot \nabla v) = \langle f, v \rangle \qquad \text{for all } v \in H_0^1, \tag{3.3}$$

which represents the weak formulation of the problem

$$-\Delta u + \lambda u = f \quad \text{in } \Omega, \quad u = 0 \quad \text{on } \partial\Omega, \quad \lambda = 1;$$

see the end of Section 2.4. We can write (3.3) in the form of (2.22) or (1.1) by defining the operator

$$\mathcal{B} : H_0^1 \to H^{-1},$$

where the value of the functional $\mathcal{B}u$ is determined for all $u \in H_0^1$ by

$$\langle \mathcal{B}u, v \rangle := \int_\Omega (uv + \nabla u \cdot \nabla v) = \langle f, v \rangle \quad \text{for all } v \in H_0^1.$$

Analogously, referring to the Friedrichs inequality (2.35), we can consider the same space $H_0^1(\Omega)$ with the equivalent inner product $(u, v)_{H_0^1} = \int_\Omega \nabla u \cdot \nabla v$. Then the associated Riesz map gives

$$\langle f, v \rangle = (\tau f, v)_V = (\tau f, v)_{H_0^1} = \int_\Omega \nabla(\tau f) \cdot \nabla v,$$

which provides with $u = \tau f$ the weak formulation of the Poisson problem

$$-\Delta u = f \quad \text{in } \Omega, \quad u = 0 \quad \text{on } \partial\Omega,$$

with

$$\mathcal{B} : H_0^1 \to H^{-1}, \quad \langle \mathcal{B}u, v \rangle = \int_\Omega \nabla u \cdot \nabla v = \langle f, v \rangle \quad \text{for all } f \in H^{-1} \text{ and for all } v \in H_0^1.$$

Summarizing, with $V = H_0^1$ and the inner product $(\cdot, \cdot)_{H^1}$,

$$\tau : H^{-1} \to H_0^1, \quad \tau f = u \in H_0^1,$$

where u is obtained by solving the problem

$$\int_\Omega (uv + \nabla u \cdot \nabla v) = \langle f, v \rangle \quad \text{for all } v \in H_0^1.$$

Analogously, with the inner product $(u, v)_{H_0^1} = \int_\Omega \nabla u \cdot \nabla v$,

$$\tau : H^{-1} \to H_0^1, \quad \tau f = u \in H_0^1,$$

where u solves

$$\int_\Omega \nabla u \cdot \nabla v = \langle f, v \rangle \quad \text{for all } v \in H_0^1.$$

These two simple cases illustrate how τ varies with the choice of the inner product in H_0^1. ∎

Different inner products determine different Riesz maps. Here we have not specified any problem $\mathscr{A} u = b$ to be solved; the Riesz map was determined solely by the choice of the inner product in V. In practice this choice is linked with the given PDE problem in order to reduce the cost of numerical computations needed for finding an appropriate numerical solution; see Chapter 4 below.

3.2 ▪ Functional equations and bilinear forms

Consider the functional equation (1.1). To each problem (1.1) expressed in terms of the bounded linear operator $\mathscr{A}$ there is an equivalent formulation expressed in terms of an associated bounded (continuous) bilinear form $a(\cdot,\cdot) : V \times V \to \mathbb{R}$. Indeed, noticing that (1.1) means

$$\text{to find } u \in V : \quad \langle \mathscr{A} u, v \rangle = \langle b, v \rangle \quad \text{for all } v \in V, \tag{3.4}$$

one can set $a(u,v) := \langle \mathscr{A} u, v \rangle$ and rewrite (3.4) as

$$\text{to find } u \in V : \quad a(u,v) = \langle b, v \rangle \quad \text{for all } v \in V. \tag{3.5}$$

Vice versa, for a given bounded bilinear form $a(\cdot,\cdot)$ fulfilling (3.5), we define $\mathscr{A} : V \to V^{\#}$ as (see also Section 2.2)

$$\mathscr{A} : u \in V \mapsto a(u,\cdot) \in V^{\#}$$

and observe that (3.5) then implies (3.4). Summarizing,

$$\langle \mathscr{A} u, v \rangle = a(u,v) \qquad \text{for all } u,v \in V, \tag{3.6}$$

where $\mathscr{A} : V \to V^{\#}$ is linear, $a(\cdot,\cdot) : V \times V \to \mathbb{R}$ is bilinear, and $\mathscr{A}$ and $a(\cdot,\cdot)$ are bounded; i.e., there is $C > 0$ such $\|\mathscr{A} u\|_{V^{\#}} \le C \|u\|_V$ for all $u \in V$, which is equivalent to

$$|a(u,v)| \le C \|u\|_V \|v\|_V \qquad \text{for all } u,v \in V.$$

Two additional requirements will be given on $\mathscr{A}$ respectively $a(\cdot,\cdot)$ in the text below. We mention them now in order to avoid misunderstandings. They can be expressed either in terms of $\mathscr{A}$ or in terms of $a(\cdot,\cdot)$, which results in a different terminology. The first one,

$$\text{there is } \alpha > 0 : \quad \langle \mathscr{A} u, u \rangle = a(u,u) \ge \alpha \|u\|_V^2, \tag{3.7}$$

says that the operator $\mathscr{A}$ is (α-)coercive[10] or $a(\cdot,\cdot)$ is V-elliptic, depending whether one likes to refer to (3.4) or rather to (3.5). The second requirement concerns the symmetry of the problem, that is,

$$\langle \mathscr{A} u, v \rangle = a(u,v) = a(v,u) = \langle \mathscr{A} v, u \rangle \quad \text{for all } u,v \in V, \tag{3.8}$$

which says that $\mathscr{A}$ is self-adjoint with respect to the duality pairing or $a(\cdot,\cdot)$ is symmetric.

Note that if $a(\cdot,\cdot)$ is bounded, V-elliptic and symmetric, then $a(\cdot,\cdot)$ is the inner product on V (topologically) equivalent to the inner product $(\cdot,\cdot)_V$.

As already mentioned in connection with the specific case in Chapter 2, symmetry (in addition to (3.7)) also enables us to investigate the problem (3.5) via techniques from the calculus of variations. Indeed, defining the functional J by

$$J(v) := \frac{1}{2} a(v,v) - \langle b, v \rangle, \qquad v \in V, \tag{3.9}$$

---

[10] In general, an operator $\mathscr{A} : V \to V^{\#}$ (not necessarily linear) is coercive if $|\langle \mathscr{A}(u), u \rangle| \to \infty$ as $\|u\|_V \to \infty$.

one observes that the existence of the minimizer of J not only implies but (in this simple framework) is equivalent to the existence of a solution of (3.5). Indeed, let $u \in V$ be a minimizer of J (i.e., $J(u) \leq J(v)$ for all $v \in V$). Then noticing that for any $v \in V$ the function $g_v : \mathbb{R} \to \mathbb{R}$ defined through

$$g_v(\zeta) := J(u + \zeta v) = \frac{1}{2}a(u,u) - \langle b,u \rangle + \zeta\left(a(u,v) - \langle b,v \rangle\right) + \frac{1}{2}\zeta^2 a(v,v)$$

is quadratic in ζ and has a minimum at $\zeta = 0$ (for any $v \in V$). Hence $g_v'(0) = 0$, which implies (3.5). Vice versa, if $u \in V$ solves (3.5), then we easily compute for any $v \in V$ that

$$\begin{aligned}
J(u + v) &= \frac{1}{2}a(u+v, u+v) - \langle b, u+v \rangle \\
&= \frac{1}{2}a(u,u) - \langle b,u \rangle + \left(a(u,v) - \langle b,v \rangle\right) + \frac{1}{2}a(v,v) \\
&= J(u) + \frac{1}{2}a(v,v) \geq J(u)
\end{aligned}$$

as $a(v,v) \geq 0$ by (3.7). Thus, $u \in V$ minimizes J over V.

We mentioned this simple standard argument here to emphasize that the concept of a weak solution to a PDE boundary-value problem such as (2.2), which is represented here by the problem (3.5), appears in the calculus of variations prior to deriving its classical formulation called the Euler or the Euler–Lagrange equation.

The derivative $g_v'(0)$ is the derivative of J at $u \in V$ in the direction $v \in V$; i.e., it is a directional derivative of J at u in the direction v and it is denoted by $\delta J(u;v)$. A functional J is said to be Gâteaux differentiable at $u \in V$ if the directional derivative $\delta J(u;v)$ exists for all $v \in V$. Then, for an arbitrary direction $v \in V$,

$$\delta J(u;v) := \lim_{\zeta \to 0} \frac{J(u + \zeta v) - J(u)}{\zeta} = g_v'(0).$$

In this context, one can sometimes find a reference to the concept of Fréchet differentiability. A functional J is said to be Fréchet differentiable at $u \in V$ if there exists a bounded linear map $DJ(u) \in V^{\#}$, called the differential of J at u, such that

$$\frac{J(u + v) - J(u) - \langle DJ(u), v \rangle}{\|v\|_V} \to 0 \quad \text{as } \|v\|_V \to 0.$$

Taking $w = \xi v$, $\|v\|_V = 1$, we get

$$\begin{aligned}
\delta J(u;v) &= \lim_{\|w\|_V \to 0} \frac{J(u + w) - J(u)}{\|w\|_V} - \langle DJ(u), v \rangle + \langle DJ(u), v \rangle \\
&= \lim_{\|w\|_V \to 0} \frac{J(u + w) - J(u) - \langle DJ(u), w \rangle}{\|w\|_V} + \langle DJ(u), v \rangle = \langle DJ(u), v \rangle.
\end{aligned}$$

Consequently, providing that the Fréchet derivative $DJ(u)$ exists, the directional derivative of J at u in the direction v exists as well and is given by $\langle DJ(u), v \rangle$.

3.3 ▪ Lax–Milgram lemma

The Lax–Milgram lemma represents an elegant tool for investigating (1.1) by using the bilinear form representation without assuming its symmetry.

Let V be a Hilbert space with the inner product $(\cdot,\cdot)_V$ and the corresponding norm $\|\cdot\|_V$, and let a bilinear form $a(\cdot,\cdot)$ be V-elliptic and bounded, i.e.,

$$\text{there is an } \alpha > 0: \qquad a(u,u) \geq \alpha\|u\|_V^2 \quad \text{for all } u \in V, \tag{3.10}$$

$$\text{there is a } C > 0: \qquad a(u,v) \leq C\|u\|_V\|v\|_V \quad \text{for all } u,v \in V; \tag{3.11}$$

then the Lax–Milgram lemma says that for each $b \in V^{\#}$ there is unique $u \in V$ so that

$$a(u,v) = \langle b,v\rangle \quad \text{for all } v \in V, \quad \text{and} \quad \|u\|_V \leq \frac{\|b\|_{V^{\#}}}{\alpha}. \tag{3.12}$$

The proof is a consequence of the Riesz representation theorem. Indeed, if $a(\cdot,\cdot)$ is, in addition, symmetric, then $a(\cdot,\cdot)$ forms an inner product in V and $\sqrt{a(\cdot,\cdot)}$ generates a norm in V that is by (3.10)–(3.11) equivalent to $\|\cdot\|_V$. Consequently, for any $b \in V^{\#}$ the Riesz representation theorem with this inner product immediately gives the first part of (3.12). The second part follows from $\alpha\|u\|_V^2 \leq a(u,u) = \langle b,u\rangle \leq \|b\|_{V^{\#}}\|u\|_V$.

In the general case, if $a(\cdot,\cdot)$ is not symmetric, then for any $u \in V$ we have $\mathscr{A}u := a(u,\cdot) \in V^{\#}$, and using the Riesz map $\tau : V^{\#} \to V$ defined by the inner product $(\cdot,\cdot)_V$ and the identifications $a(u,v) = (\tau\mathscr{A}u,v)_V$ and $\langle b,v\rangle = (\tau b,v)_V$ for all $v \in V$ we get the equivalence

$$a(u,v) = \langle b,v\rangle \text{ for all } v \in V \iff (\tau\mathscr{A}u,v)_V = (\tau b,v)_V \text{ for all } v \in V. \tag{3.13}$$

The proof is completed by showing that the (composed) linear mapping $\tau\mathscr{A} : V \to V$ is one-to-one and onto. See, e.g., [62] or the original paper [121] for details and [126] for a constructive proof based on the Banach fixed point theorem.

Note that (3.10) and (3.11) can be replaced by the equivalent conditions

$$\text{there is an } \alpha > 0: \qquad \inf_{u\in V;\|u\|_V=1} a(u,u) \geq \alpha, \tag{3.14}$$

$$\text{there is a } C > 0: \qquad \sup_{u,v\in V;\|u\|_V=\|v\|_V=1} a(u,v) \leq C, \tag{3.15}$$

which in terms of $\mathscr{A}$ means that

$$\text{there is an } \alpha > 0: \qquad \inf_{u\in V;\|u\|_V=1} \langle\mathscr{A}u,u\rangle \geq \alpha, \tag{3.16}$$

$$\text{there is a } C > 0: \qquad \sup_{u\in V;\|u\|_V=1} \|\mathscr{A}u\|_{V^{\#}} \leq C. \tag{3.17}$$

In the subsequent text it is convenient to state (3.16) and (3.17) (or (3.10) and (3.11)) with the best possible α and C, denoted by α_* and C_*, respectively. Clearly, the smallest C_* is defined through

$$C_* := \sup_{u\in V;\|u\|_V=1} \|\mathscr{A}u\|_{V^{\#}}.$$

Concerning the largest α_*, note that (3.12) implies that for any $f \in V^{\#}$,

$$\|\mathscr{A}^{-1}f\|_V \leq \frac{\|f\|_{V^{\#}}}{\alpha}. \tag{3.18}$$

Consequently, setting

$$\alpha_* := \left(\sup_{f\in V^{\#};\|f\|_{V^{\#}}=1} \|\mathscr{A}^{-1}f\|_V\right)^{-1},$$

we observe that $\alpha^* \geq \alpha$ for any α fulfilling (3.16). This optimal choice, which can be equivalently expressed as

$$\alpha_* = \inf_{u \in V;\, \|u\|_V = 1} \langle \mathscr{A} u, u \rangle, \tag{3.19}$$

leads (instead of (3.10) and (3.11)) to the conditions

$$a(u,u) \geq \alpha_* \|u\|_V^2 \quad \text{for all } u \in V \quad \text{with } \alpha_*^{-1} := \sup_{b \in V^\#;\, \|b\|_{V^\#} = 1} \|\mathscr{A}^{-1} b\|_V, \tag{3.20}$$

$$a(u,v) \leq C_* \|u\|_V \|v\|_V \quad \text{for all } u,v \in V \quad \text{with } C_* := \sup_{u \in V;\, \|u\|_V = 1} \|\mathscr{A} u\|_{V^\#}. \tag{3.21}$$

With the obvious definition of the operator norm (which we use here just to express the following relations in a compact form), (3.20) and (3.21) imply

$$\|\mathscr{A}^{-1}\|_{\mathscr{L}(V^\#, V)}^{-1} \|u\|_V^2 \leq a(u,u) \leq \|\mathscr{A}\|_{\mathscr{L}(V, V^\#)} \|u\|_V^2 \quad \text{for all } u \in V. \tag{3.22}$$

Chapter 4

Riesz Map and Operator Preconditioning

Until now we were concerned with the *analysis* of the PDE problem (1.1). In this chapter the choice of the inner product $(\cdot,\cdot)_V$ and the associated Riesz map $\tau : V^\# \to V$ is linked with *transformation* of the functional equation (1.1) into a form which is more suitable for efficient numerical solution.

Consider the problem (1.1) (or, equivalently, (3.4) or (3.5)), i.e.,

$$\mathscr{A} u = b, \quad \mathscr{A} : V \to V^\#, \quad u \in V, \quad b \in V^\#, \tag{4.1}$$

and an inner product $(\cdot,\cdot)_V : V \times V \to \mathbb{R}$. The Riesz map $\tau : V^\# \to V$ defined for $f \in V^\#$ by $(\tau f, v)_V := \langle f, v \rangle$ for all $v \in V$ is determined by the choice of the inner product on V. The formulation (3.5) rewritten as

$$\langle \mathscr{A} u - b, v \rangle = 0 \quad \text{for all } v \in V$$

(the weak formulation of the PDE problem) can be equivalently formulated using the Riesz map as

$$(\tau(\mathscr{A} u) - \tau b, v)_V = 0 \quad \text{for all } v \in V.$$

Clearly, the Riesz map τ can be interpreted as a *transformation* of the original problem $\mathscr{A} u = f$ in $V^\#$ into the equation in the solution space V,

$$\tau \mathscr{A} u = \tau b, \quad \tau \mathscr{A} : V \to V, \quad u \in V, \quad \tau b \in V. \tag{4.2}$$

This transformation is commonly and inaccurately called *operator preconditioning*, and it motivates the construction of acceleration techniques used in order to improve the behavior of iterative methods for solving the associated discretized problems. The relationship between discretization and preconditioning is further studied in Chapter 8.

Let the bilinear form $a(\cdot,\cdot)$ be symmetric, bounded, and V-elliptic. Then it defines the inner product $(\cdot,\cdot)_a$ on V, $(u,v)_a := a(u,v)$ for all $u, v \in V$. Consider the Riesz map τ_a defined by the inner product $(\cdot,\cdot)_a$. For the functional $\mathscr{A} u \in V^\#$ and the inner product $(\cdot,\cdot)_a$ on V we get

$$\langle \mathscr{A} u, v \rangle = (\tau_a(\mathscr{A} u), v)_a$$

and from the definition of $\mathscr{A}$ (see (3.6))

$$\langle \mathscr{A} u, v \rangle = (u, v)_a,$$

both of which hold for all $v \in V$. Then

$$\tau_a(\mathscr{A}u) = u, \quad \text{i.e,} \quad \tau_a = \mathscr{A}^{-1} : V^\# \to V.$$

Summarizing, with

$$(u, v)_V := (u, v)_a := a(u, v) = \langle \mathscr{A}u, v \rangle,$$

the Riesz map τ is nothing but the inverse of the operator $\mathscr{A}$ representing the bilinear form $a(\cdot, \cdot)$, $\tau = \mathscr{A}^{-1}$, and $\tau_a \mathscr{A}u = \tau_a b$ will reduce to $u = \mathscr{A}^{-1}b$. Considering this, the motivation of the operator preconditioning seems to be to take the inner product $(\cdot, \cdot)_V$ as close as possible to the bilinear form $a(u, v)$ associated with $\mathscr{A}$. The matter is, however, not that simple. In numerical computations one has to take into account the cost at which a numerical approximation to the solution u is computed. The better the approximation of $a(u, v)$ by $(\cdot, \cdot)_V$ is, the more expensive, in general, the construction and application of the Riesz map τ are. In order to minimize the overall computational costs, the gain in acceleration of convergence in iterative computations must more than pay off for the extra costs associated with the preconditioner.

The crucial point is that in the operator preconditioning based on the Riesz map technique we are dealing with a *transformation of the original PDE model into a new form*, which ideally is, after discretization, computationally more feasible while preserving (at the operator level) the information contained in the model. However, the computational costs can be evaluated only in terms of the discretized problem and the application of the particular iterative method. This in practice involves also issues such as efficient implementations of numerical algorithms on the given (parallel) computer architectures, which assumes also the analysis of possible effects of finite precision arithmetic (rounding errors). The prevailing numerical PDEs approaches to estimating computational cost are based on a substantial simplification using a priori bounds; see Chapters 8, 9, and 11.

The idea of operator preconditioning goes back at least to the early works and Ph.D. thesis of Klawonn [104, 105] and to the independent works of Arnold, Falk, and Winther [7, 8]; see also various developments presented in [177, 133, 43, 42, 154]. Construction of block diagonal preconditioners for discretized saddle-point problems represented one of the main motivations which led to understanding the preconditioner as representation of an appropriate inner product in the sense described above. The applications range from elasticity [108] to fluid mechanics, where the interplay of operator preconditioning and algebraic preconditioning has been used for many years by Elman, Silvester, and Wathen and their collaborators and students; see, e.g., [175], the book [59], and [181, 58, 180, 150, 149]. The topic has also been surveyed in the papers [130, 103, 204, 12, 98]. Many of the ideas presented in the later works are related to the concept of equivalent operators comprehensively presented in the remarkable paper by Faber, Manteuffel, and Parter [63]. The authors refer to the early papers by D'yakonov [55, 56] and Gunn [90, 91], where the inverse of the Laplace operator was used as a preconditioner and the spectral equivalence concept was used for proving bounds independent of the mesh. The paper [63] also points out some issues which can limit applicability of the developed concept. In particular, the *quantitative* reasoning in the evaluation of operator preconditioning (i.e., of the choice of the inner product $(\cdot, \cdot)_V$) or in using the concept of equivalent operators in these publications, as a rule, focuses on the construction of bounds using condition numbers.

For a bounded and α-coercive linear operator $\mathscr{A} : V \to V^\#$, the condition number is defined as (see (3.22))

$$\varkappa(\mathscr{A}) := \sup_{\|u\|_V = 1} \|\mathscr{A}u\|_{V^\#} \ \sup_{\|z\|_{V^\#} = 1} \|\mathscr{A}^{-1}z\|_V. \tag{4.3}$$

By (3.17) and (3.18) we immediately observe that

$$\sup_{\|u\|_V=1} \|\mathscr{A}u\|_{V^\#} \leq C \quad \text{and} \quad \sup_{\|z\|_{V^\#}=1} \|\mathscr{A}^{-1}z\|_V \leq 1/\alpha.$$

With

$$\sup_{\|z\|_{V^\#}=1} \|\mathscr{A}^{-1}z\|_V = \sup_{z \in V^\#} \frac{\|\mathscr{A}^{-1}z\|_V}{\|z\|_{V^\#}} = \sup_{w \in V} \frac{\|w\|_V}{\|\mathscr{A}w\|_{V^\#}} = \sup_{\|w\|_V=1} \frac{1}{\|\mathscr{A}w\|_{V^\#}}$$

and considering

$$\|\mathscr{A}w\|_{V^\#} = \sup_{\|q\|_V=1} \langle \mathscr{A}w,q \rangle \geq \langle \mathscr{A}w,w \rangle \quad \text{for } \|w\|_V = 1$$

we get

$$\|\mathscr{A}w\|_{V^\#} \geq \inf_{\|w\|_V=1} \langle \mathscr{A}w,w \rangle$$

and

$$\sup_{\|z\|_{V^\#}=1} \|\mathscr{A}^{-1}z\|_V \leq \frac{1}{\inf_{\|w\|_V=1}\langle \mathscr{A}w,w \rangle}. \tag{4.4}$$

Therefore it follows from (4.3) and (4.4) that

$$x(\mathscr{A}) \leq \frac{C}{\alpha} \quad \text{and} \quad x(\mathscr{A}) \leq \frac{\sup_{\|u\|_V=1}\|\mathscr{A}u\|_{V^\#}}{\inf_{\|u\|_V=1}\langle \mathscr{A}u,u \rangle}. \tag{4.5}$$

Using (3.1) and (3.2) we finally conclude that

$$x(\mathscr{A}) \leq \frac{\sup_{\|u\|_V=1}\|\tau\mathscr{A}u\|_V}{\inf_{\|u\|_V=1}(\tau\mathscr{A}u,u)_V}. \tag{4.6}$$

This suggests that $x(\mathscr{A})$ and the bound (4.6) is indeed related to the (operator preconditioned) formulation (4.2) of the problem in the solution space V.

Alternatively, starting from the definition (4.3) and using (3.20), (3.21) together with (3.19) we directly obtain (instead of (4.5))

$$x(\mathscr{A}) = \frac{C_*}{\alpha_*} \quad \text{and} \quad x(\mathscr{A}) = \frac{\sup_{\|u\|_V=1}\|\mathscr{A}u\|_{V^\#}}{\inf_{\|u\|_V=1}\langle \mathscr{A}u,u \rangle}, \tag{4.7}$$

which as above leads to the improvement over (4.6):

$$x(\mathscr{A}) = \frac{\sup_{\|u\|_V=1}\|\tau\mathscr{A}u\|_V}{\inf_{\|u\|_V=1}(\tau\mathscr{A}u,u)_V}. \tag{4.8}$$

In order to perform numerical computations, the problem must be discretized. The operator preconditioning approach, as presented in the literature, then aims at proving that (4.6) provides uniform bounds on the condition numbers of the discretized preconditioned operators (and, consequently, on their algebraic matrix representations) independently on the discretization (and, possibly also on some problem related) parameters; see, e.g., [12, 103, 130]. The beauty of this idea is expressed by the following quote from the paper by Zulehner [204]:

> *Knowledge of robust estimates not only contributes to the question of well-posedness, but also to discretization error estimates and the construction of efficient solvers for the discretized problems. In the discretized case, having robust estimates for an inner product translates to having a block diagonal preconditioner for the linear operator with robust estimates for the condition number. This would immediately imply that Krylov subspace methods like the minimal residual method converge with convergence rates independent of h.*

This quote shows how the operator preconditioning approach described above links the functional analysis-based description of the infinite-dimensional operators with the numerical analysis of the discretization (a priori error bounds and convergence of discretization schemes) and even with the analysis of convergence of Krylov subspace methods for solving discretized systems of linear algebraic equations. This generality and focus on a priori bounds also has drawbacks, however. As stated by Hiptmair in [98],

> *Operator preconditioning is a very general recipe for constructing preconditioners for the operators arising from discrete BVPs and integral equations. It is simple to apply, but may not be particularly efficient, because, in case the bound is large, the operator preconditioning policy offers no hint how to improve the preconditioner. Hence, operator preconditioning may often achieve the much-vaunted mesh independence of the preconditioner, but it may not perform satisfactorily on a given mesh.*

Similarly, Faber, Manteuffel, and Parter state in relation to the associated concept of equivalent operators already in 1990 [63, pp. 114–115]:

> *For a fixed h [which means the discretization parameter], using a preconditioning strategy based upon an equivalent operator may not be superior to classical methods .... Equivalence alone is not sufficient for a good preconditioning strategy. One must also choose an equivalent operator for which the bound ... is small. The above observations indicate that a more precise measure of the "closeness" of two operators is required to evaluate preconditioning strategies.*

They also address the point which is, in our context of Krylov subspace methods, of particular importance (see Chapter 11):

> *There is no flaw in the analysis, only a flaw in the conclusion drawn from the analysis. These bounds are only asymptotic .... Methods with similar asymptotic work estimates may behave quite differently in practice.*

In operator preconditioning (as well as in most techniques based on domain decomposition ideas, multilevel approaches, hierarchical bases and wavelet discretizations), the link between infinite-dimensional description, discretization, and computation is made, as illustrated above, at a price of nontrivial simplifications reducing the investigation of convergence rate of iterative methods to various formulas based on condition numbers. We will further examine these issues using CG as the main representative of Krylov subspace methods. CG is well suited for solving discretized problems with linear elliptic, bounded, coercive, and self-adjoint operators.

As $\tau \mathscr{A} : V \to V$, we can form for $g \in V$ the Krylov sequence $g, \tau \mathscr{A} g, (\tau \mathscr{A})^2 g, \dots$ in V and define Krylov subspace methods in infinite-dimensional Hilbert spaces. Since Krylov subspace methods are based on taking powers of the operator *with respect to a given vector*, this requires that the operator maps V to V, which highlights the role of the Riesz map. Following the comprehensive description in [92], we will illustrate this on derivation of CG; see also [130] and [74, Section 16], where CG is formulated using the bilinear form representation (3.5). We begin in the next chapter with the infinite-dimensional CG and investigate various aspects of discretization in further chapters.

Chapter 5

Conjugate Gradient Method in Hilbert Spaces

Using the notation of Sections 3.2 and 3.3 and of Chapter 4, consider the self-adjoint, bounded, and α-coercive linear operator $\mathscr{A} : V \to V^{\#}$, the associated bilinear form

$$a(u,v) := \langle \mathscr{A} u, v \rangle : V \times V \to \mathbb{R},$$

the inner product $(u,v)_a := a(u,v)$ and the induced norm $\| \cdot \|_a$ on V. Our goal is to construct a method for solving the operator equation (4.1), where we use for consistency with the standard algebraic description the unknown x instead of u, giving

$$\mathscr{A} x = b, \quad x \in V, \quad b \in V^{\#}. \tag{5.1}$$

The approximations x_n to the solution x are for $n = 1, 2, \ldots$ sought using the subspaces

$$\begin{aligned}
x_n &\in x_0 + K_n(\tau \mathscr{A}, \tau r_0) \\
&:= x_0 + \operatorname{span}\{\tau r_0, \tau \mathscr{A}(\tau r_0), (\tau \mathscr{A})^2(\tau r_0), \ldots, (\tau \mathscr{A})^{n-1}(\tau r_0)\}
\end{aligned} \tag{5.2}$$

defined by $\mathscr{A}$, $r_0 = b - \mathscr{A} x_0$ with a properly chosen[11] initial approximation $x_0 \in V$ and the Riesz map τ associated with the properly chosen inner product $(\cdot, \cdot)_V : V \times V$. This inner product is typically different from $(\cdot, \cdot)_a$; the choice $(\cdot, \cdot)_V = (\cdot, \cdot)_a$ would mean inverting the operator $\mathscr{A}$. On the other hand, $(\cdot, \cdot)_V$ should be close enough to $(\cdot, \cdot)_a$ to ensure good approximation properties of $K_n := K_n(\tau \mathscr{A}, \tau r_0)$; see also the discussion in Chapter 4.

In many physical applications (see Chapter 2 for an example) it makes good sense to minimize the $\| \cdot \|_a$ norm of the error $x - x_n$, i.e., to look for x_n such that

$$\|x - x_n\|_a = \min_{z \in x_0 + K_n} \|x - z\|_a. \tag{5.3}$$

In order to reformulate (5.3), consider the orthogonal decomposition of the initial error

$$x - x_0 = (x - x_0)|_{K_n} + (x - x_0)|_{K_n^{\perp_a}}, \tag{5.4}$$

where $(x - x_0)|_{K_n} \in K_n$ and $(x - x_0)|_{K_n^{\perp_a}}$ is its orthogonal complement with respect to $(\cdot, \cdot)_a$, i.e.,

$$\left((x - x_0)|_{K_n}, (x - x_0)|_{K_n^{\perp_a}}\right)_a = \left\langle \mathscr{A}(x - x_0)|_{K_n}, (x - x_0)|_{K_n^{\perp_a}} \right\rangle = 0.$$

---

[11] If no information guiding the choice of x_0 is available, then one should take $x_0 = 0$.

Since $x_n - x_0 \in K_n$, it is worth comparing (5.4) with the simple equality

$$x - x_0 = (x_n - x_0) + (x - x_n),$$

which gives

$$x - x_n = (x - x_0) - (x_n - x_0) = \underbrace{(x - x_0)|_{K_n} - (x_n - x_0)}_{\in K_n} + \underbrace{(x - x_0)|_{K_n^{\perp_a}}}_{\in K_n^{\perp_a}}.$$

Clearly, the minimum (5.3) is achieved if and only if

$$x_n - x_0 = (x - x_0)|_{K_n}, \quad x - x_n = (x - x_0)|_{K_n^{\perp_a}} \tag{5.5}$$

or, equivalently, if and only if

$$(x - x_n) \perp_a K_n \quad \text{i.e.,} \quad \langle \mathscr{A}(x - x_n), w \rangle = 0 \quad \text{for all } w \in K_n. \tag{5.6}$$

Using $\mathscr{A}(x - x_n) = b - \mathscr{A}x_n = r_n$, (5.6) gives

$$\langle r_n, w \rangle = (\tau r_n, w)_V = 0 \quad \text{for all } w \in K_n, \tag{5.7}$$

which is nothing but the form of the Galerkin orthogonality condition with respect to the nth Krylov subspace K_n.

Remark 2. As we will see in more detail in Section 5.2, the Galerkin orthogonality (5.7) determines the n-dimensional problem which can be viewed as the discretization of $\tau \mathscr{A} x = \tau b$ by the orthogonal projection onto K_n. If there is an efficient way of performing the corresponding (operator) algebraic operations in V and $V^{\#}$, the CG approximation x_n would provide the n-dimensional approximation to x minimizing the energy norm of the error over the Krylov subspace K_n.

5.1 ▪ Algorithmic construction of CG

In order to derive the computational algorithm satisfying the equivalent conditions (5.3), (5.6), and (5.7), we consider the quadratic functional (see (3.9) in Chapter 3 and (2.23) in Chapter 2)

$$J(z) = \frac{1}{2} \langle \mathscr{A} z, z \rangle - \langle b, z \rangle, \quad z \in V, \tag{5.8}$$

and the minimization problem

$$\min_{z \in V} J(z). \tag{5.9}$$

Under the assumptions of boundedness, α-coercivity, and self-adjointness above, this problem has a unique solution x given by (5.1); see Section 3.2. Moreover, since

$$J(z) = \frac{1}{2} \langle \mathscr{A}(z - x), z - x \rangle + J(x) = \frac{1}{2} \|x - z\|_a^2 + J(x), \tag{5.10}$$

the minimum in (5.3) measured in the $\|\cdot\|_a$ norm is attained at x_n, which at the same time minimizes $J(z)$ over the affine subspace $x_0 + K_n$.

Remark 3. The equality (5.10) is often written in a different form. Using

$$0 = \langle b - \mathscr{A}x, x \rangle = \langle b, x \rangle - \langle \mathscr{A}x, x \rangle = -J(x) - \frac{1}{2} \langle \mathscr{A}x, x \rangle$$

we get from (5.10)

$$\frac{1}{2}\|z-x\|_a^2 = \frac{1}{2}\langle \mathscr{A}(z-x), z-x\rangle = J(z) + \frac{1}{2}\langle \mathscr{A}x, x\rangle.$$

Summarizing, our starting point for the construction of the computational algorithm is to look for an approximation $x_n \in x_0 + K_n$ to the solution x of (5.1) such that it gives the minimum in (5.3). This can be constructed by finding the minimum of the functional J introduced in (5.8) over $x_0 + K_n$, i.e.,

$$J(x_n) = \min_{z \in x_0 + K_n} J(z), \tag{5.11}$$

which is uniquely determined by the condition (5.7). Thus, we have

$$\|x - x_n\|_a = \min_{z \in x_0 + K_n} \|x - z\|_a \quad \text{or} \quad x_n = \operatorname*{argmin}_{z \in x_0 + K_n} J(z). \tag{5.12}$$

We will construct a sequence of direction vectors $p_0, p_1, \ldots$ and scalars $\alpha_0, \alpha_1, \ldots$ such that, starting from x_0,

$$x_n = x_{n-1} + \alpha_{n-1} p_{n-1}, \quad n = 1, 2, \ldots,$$

where $p_{n-1} \in V$ and α_{n-1} ensures the minimization of the functional J along the line $z(\alpha) = x_{n-1} + \alpha p_{n-1}$. This gives, with $r_n := b - \mathscr{A}x_n$,

$$\alpha_{n-1} = \frac{\langle r_{n-1}, p_{n-1}\rangle}{\langle \mathscr{A}p_{n-1}, p_{n-1}\rangle} \tag{5.13}$$

and, as an immediate consequence, the orthogonality of the nth residual to the direction vector p_{n-1} with respect to the duality pairing $\langle \cdot, \cdot\rangle$,

$$\langle r_n, p_{n-1}\rangle = \langle b - \mathscr{A}x_n, p_{n-1}\rangle = \langle r_{n-1}, p_{n-1}\rangle - \alpha_{n-1}\langle \mathscr{A}p_{n-1}, p_{n-1}\rangle = 0.$$

The key to the derivation of the algorithm is in the choice of the direction vectors. For $n = 1$ we simply take p_0 as the direction of the steepest descent (see below for the explicit specification). In order to guarantee that for $n = 2$ the minimization of $J(z_1)$ along the line $z_1(\alpha) = x_1 + \alpha p_1$ ensures at the same time the minimization of $J(z)$ over the affine subspace determined by the two-dimensional subspace generated by the vectors p_0 and p_1, we must use the direction vector p_1 such that $p_1 \perp_a p_0$. In general, if we satisfy

$$K_n := K_n(\tau \mathscr{A}, \tau r_0) = \operatorname{span}\{p_0, p_1, \ldots, p_{n-1}\} \quad \text{and} \quad p_i \perp_a p_j, \ i \neq j, \tag{5.14}$$

then the one-dimensional line minimizations of $J(z_\ell)$ along the given n individual lines $z_\ell(\alpha) = x_{\ell-1} + \alpha p_{\ell-1}, \ell = 1, 2, \ldots, n$, will be equivalent to minimization of the functional $J(z)$ over the whole n-dimensional affine subspace $x_0 + K_n$, i.e., it will guarantee that x_n satisfies (5.12). Indeed, provided that (5.14) holds,

$$x - x_0 = \sum_{\ell=0}^{n-1} \alpha_\ell p_\ell + (x - x_n)$$

represents the orthogonal expansion of the initial error $x - x_0$ along the direction vectors (also called the search vectors) $p_0, \ldots, p_{n-1}$ using the inner product $(\cdot, \cdot)_a$ defined by the bilinear form $a(\cdot,\cdot)$. Therefore for the remainder we must have $(x - x_n) \perp_a K_n$; see (5.3)–(5.7).

The last paragraph gives the essence of CG viewed as minimization of the quadratic functional or, equivalently, as minimization of the energy norm of the error, over the affine subspaces $x_0 + K_\ell$, $\ell = 1, 2, \ldots, n$. The crucial point of the special choice of the direction vectors which guarantees replacing the n one-dimensional minimizations (such as in the method of steepest descent) by the *single n-dimensional minimization while preserving formally the step-by-step updating structure of the algorithm* is very elegant. All subsequent technical steps are nothing but its logical and straightforward technical realization.

Here we emphasize that CG for solving the linear problem (5.1) via minimizing the quadratic functional (5.11) *should not be viewed as a simplification* of CG known from nonlinear optimization; see, e.g., [49, 127, 140]. It must be emphasized that CG for solving the *linear* problem (5.1) is a highly *nonlinear method*, with the nonlinearity in $\mathscr{A}$ and b emerging as a straightforward consequence of the minimization over the Krylov subspaces $K_\ell(\tau\mathscr{A}, \tau r_0)$, $\ell = 1, 2 \ldots$; for more details see Section 5.2 and also Chapter 11.

The rest of the derivation is a technical exercise with one subtle point. It is important to realize that the steepest descent direction in minimizing $J(z)$ *depends on the inner product* $(\cdot,\cdot)_V$ *considered in* V. In order to avoid confusion, recall that the Hilbert space V is equipped with the inner product $(\cdot,\cdot)_V$ associated with the Riesz map τ and subsequently also with the Krylov subspaces $K_n(\tau\mathscr{A}, \tau r_0)$. As stated above, the inner product $(\cdot,\cdot)_V$ is, in general, different from the inner product $(\cdot,\cdot)_a$ which determines the required minimization property (5.3) of the desired CG method. The steepest descent direction d_ℓ at the point x_ℓ minimizes the directional derivative; i.e., it gives the minimum of (see (5.10) and also Section 3.2)

$$\delta J(x_\ell; d_\ell) = \lim_{v \to 0} \frac{J(x_\ell + vd_\ell) - J(x_\ell)}{v} = \langle \mathscr{A} x_\ell - b, d_\ell \rangle = -\langle r_\ell, d_\ell \rangle.$$

Here $r_\ell = b - \mathscr{A} x_\ell \in V^\#$ while $d_\ell \in V$, $\|d_\ell\|_V = 1$. In order to minimize the directional derivative (i.e., to maximize its absolute value), consider the Riesz map τ associated with the inner product $(\cdot,\cdot)_V$ on V, which immediately gives

$$-\langle r_\ell, d_\ell \rangle = -(\tau r_\ell, d_\ell)_V,$$

and the minimum is attained for

$$d_\ell = \tau r_\ell / \|\tau r_\ell\|_V. \tag{5.15}$$

As mentioned above, the first direction vector is taken as the direction of the steepest descent, $p_0 = \tau r_0$ (in order to minimize the computational costs of individual iterations, it is convenient to skip the normalization of the direction vectors). The second direction vector $p_1 \in K_2(\tau\mathscr{A}, \tau r_0)$ can be generated as a linear combination $p_1 = \tau r_1 + \beta_1 p_0$, where $r_1 = b - \mathscr{A} x_1 \in V^\#$ and $\tau r_1 \in K_2(\tau\mathscr{A}, \tau r_0)$. The scalar β_1 is given by the orthogonality condition $p_1 \perp_a p_0$. This inspires an attempt for the general step

$$p_n = \tau r_n + \beta_n p_{n-1},$$

with the condition $p_n \perp_a p_{n-1}$ giving

$$\beta_n = -\frac{\langle \mathscr{A} p_{n-1}, \tau r_n \rangle}{\langle \mathscr{A} p_{n-1}, p_{n-1} \rangle}.$$

Summarizing, the considerations above determine the following algorithm for computing approximations $x_1, x_2, \ldots$ to the solution x of the equation (5.1):

ALGORITHM 5.1.

CG (preliminary version)

Initialize: Given $\mathscr{A} : V \to V^{\#}$ self-adjoint, bounded, and coercive, $b \in V^{\#}$, an initial approximation $x_0 \in V$, the duality pairing $\langle \cdot, \cdot \rangle : V^{\#} \times V \to \mathbb{R}$, and the inner product $(\cdot, \cdot)_V : V \times V \to \mathbb{R}$ determining the Riesz map τ,

$$\text{set } r_0 = b - \mathscr{A} x_0 \in V^{\#}, \quad p_0 = \tau r_0 \in V.$$

for $n = 1, 2, \ldots, n_{\max}$ **do**

$$\alpha_{n-1} = \frac{\langle r_{n-1}, p_{n-1} \rangle}{\langle \mathscr{A} p_{n-1}, p_{n-1} \rangle} \qquad \triangleright \text{(see the modification below)}$$

$$x_n = x_{n-1} + \alpha_{n-1} p_{n-1} \qquad \triangleright \text{stop when the stopping criterion is satisfied}$$

$$r_n = r_{n-1} - \alpha_{n-1} \mathscr{A} p_{n-1} \qquad \triangleright (= b - \mathscr{A} x_n)$$

$$\beta_n = -\frac{\langle \mathscr{A} p_{n-1}, \tau r_n \rangle}{\langle \mathscr{A} p_{n-1}, p_{n-1} \rangle} \qquad \triangleright \text{(see the modification below)}$$

$$p_n = \tau r_n + \beta_n p_{n-1}$$

end for

The stopping criterion is problem dependent and should be based on a posteriori evaluation of the size of the error $x - x_n$ (this is by no means a simple task even in a purely algebraic setting; see, e.g., [3, Section 4], [124, Chapter 5], [147], and the references given there, with more discussion on this topic in Chapters 11 and 12 later). This fact makes any a priori quantification of the computational costs and comparison of different approaches *without considering a particular problem including particular input data* very difficult; see [124, Section 5.4]. The maximal number of iterations $n_{\max}$ is given as a safety measure for the case when the stopping criterion is too tight to be reached within an acceptable computational time.

We now prove by induction that the presented *local orthogonality* properties of the direction vectors $p_\ell \perp_a p_{\ell-1}$ and the choice $\alpha_{\ell-1}$ in the steps $\ell = 1, \ldots, n$ ensure the *global orthogonality* relations

$$\begin{aligned}
\langle r_i, \tau r_j \rangle = (\tau r_i, \tau r_j)_V &= 0 \\
(p_i, p_j)_a = \langle \mathscr{A} p_i, p_j \rangle &= 0 \\
\langle r_i, p_j \rangle &= 0
\end{aligned} \qquad \begin{aligned} &\text{for all } i = 0, 1, \ldots, n, \\ &\text{for all } j = 0, 1, \ldots, i - 1. \end{aligned} \qquad (5.16)$$

Although the proof represents a simple technical exercise, we include it here for completeness. (Even in the finite-dimensional matrix computation setting the proof is rarely seen in the literature.) Since for any $\ell = 1, 2, \ldots$, by the choice of $\alpha_{\ell-1}$, the residual r_ℓ is orthogonal (with respect to the duality pairing) to the direction vector $p_{\ell-1}$ from the previous step, $\langle r_1, \tau r_0 \rangle = \langle r_1, p_0 \rangle = 0$ and (5.16) holds for $n = 1$. The equality $\langle \mathscr{A} p_1, p_0 \rangle = 0$ holds by construction.

Let (5.16) hold, by induction assumption, for $n - 1$. Since $(p_n, p_{n-1})_a = 0$ and also $\langle r_n, p_{n-1} \rangle = 0$ by construction, we need to verify that $(p_n, p_j)_a = 0$ and $\langle r_n, p_j \rangle = 0$ for $j = 0, \ldots, n - 2$ and $\langle r_n, \tau r_j \rangle = 0$ for $j = 0, \ldots, n - 1$.

In this direction, we first notice that simple manipulations give, for $j < n - 1$,

$$
\begin{aligned}
(p_n, p_j)_a &= \langle \mathscr{A} p_n, p_j \rangle = \langle \mathscr{A}(\tau r_n), p_j \rangle + \beta_n \langle \mathscr{A} p_{n-1}, p_j \rangle = \langle \mathscr{A}(\tau r_n), p_j \rangle \\
&= \langle \mathscr{A} p_j, \tau r_n \rangle = \frac{1}{\alpha_j} \langle r_j - r_{j+1}, \tau r_n \rangle \\
&= \frac{1}{\alpha_j} \langle r_n, \tau r_j - \tau r_{j+1} \rangle = \frac{1}{\alpha_j} (\tau r_n, \tau r_j - \tau r_{j+1})_V,
\end{aligned}
\tag{5.17}
$$

which means that the orthogonality of the direction vectors with respect to the inner product $(\cdot, \cdot)_a$ immediately follows from the orthogonality of the residuals transformed by the Riesz map with respect to the inner product $(\cdot, \cdot)_V$. Here we have used the fact that, for any $g \in V^\#$, $h \in V^\#$,

$$
\langle g, \tau h \rangle = (\tau g, \tau h)_V = (\tau h, \tau g)_V = \langle h, \tau g \rangle.
$$

Next, using the formula for direction vectors from Algorithm 5.1 (and by construction of p_{n-1}) we get $\langle \mathscr{A} p_{n-1}, \tau r_{n-1} \rangle = \langle \mathscr{A} p_{n-1}, p_{n-1} \rangle$. Therefore

$$
\begin{aligned}
(\tau r_n, \tau r_{n-1})_V = \langle r_n, \tau r_{n-1} \rangle &= \langle r_{n-1}, \tau r_{n-1} \rangle - \alpha_{n-1} \langle \mathscr{A} p_{n-1}, \tau r_{n-1} \rangle \\
&= \langle r_{n-1}, p_{n-1} \rangle - \beta_{n-1} \langle r_{n-1}, p_{n-2} \rangle - \alpha_{n-1} \langle \mathscr{A} p_{n-1}, p_{n-1} \rangle \\
&= \langle r_{n-1}, p_{n-1} \rangle - \alpha_{n-1} \langle \mathscr{A} p_{n-1}, p_{n-1} \rangle \\
&= 0,
\end{aligned}
$$

where the last equation is due to the choice of α_{n-1} in Algorithm 5.1. For $j < n - 1$, using analogous substitutions as above and the induction assumption,

$$
\begin{aligned}
\langle r_n, \tau r_j \rangle &= \langle r_{n-1}, \tau r_j \rangle - \alpha_{n-1} \langle \mathscr{A} p_{n-1}, \tau r_j \rangle = -\alpha_{n-1} \langle \mathscr{A} p_{n-1}, \tau r_j \rangle \\
&= -\alpha_{n-1} \langle \mathscr{A} p_{n-1}, p_j - \beta_j p_{j-1} \rangle = 0,
\end{aligned}
$$

where for $j = 0$ the term $\beta_0 p_{-1}$ is nonexistent. Hence the first part and, owing to (5.17), also the second part of (5.16) are proved. Finally, for $j > 0$,

$$
\begin{aligned}
\langle r_n, p_j \rangle &= \langle r_n, \tau r_j \rangle + \beta_j \langle r_n, p_{j-1} \rangle \\
&= \beta_j (\langle r_{n-1}, p_{j-1} \rangle - \alpha_{j-1} \langle \mathscr{A} p_{n-1}, p_{j-1} \rangle) = 0,
\end{aligned}
$$

and $\langle r_n, p_0 \rangle = \langle r_n, \tau r_0 \rangle = 0$ has already been proved.

The orthogonality relations (5.16) allow us to modify the coefficients α_{n-1} and β_n. Indeed,

$$
\langle r_{n-1}, p_{n-1} \rangle = \langle r_{n-1}, \tau r_{n-1} \rangle + \beta_{n-1} \langle r_{n-1}, p_{n-2} \rangle = \langle r_{n-1}, \tau r_{n-1} \rangle
$$

and therefore

$$
\alpha_{n-1} = \frac{\langle r_{n-1}, \tau r_{n-1} \rangle}{\langle \mathscr{A} p_{n-1}, p_{n-1} \rangle}.
\tag{5.18}
$$

In addition to that,

$$\beta_n = -\frac{\langle \mathscr{A} p_{n-1}, \tau r_n \rangle}{\langle \mathscr{A} p_{n-1}, p_{n-1} \rangle} = \frac{1}{\alpha_{n-1}} \left(\frac{\langle r_n, \tau r_n \rangle}{\langle \mathscr{A} p_{n-1}, p_{n-1} \rangle} - \frac{\langle r_{n-1}, \tau r_n \rangle}{\langle \mathscr{A} p_{n-1}, p_{n-1} \rangle} \right)$$
$$= \frac{1}{\alpha_{n-1}} \frac{\langle r_n, \tau r_n \rangle}{\langle \mathscr{A} p_{n-1}, p_{n-1} \rangle} = \frac{\langle r_n, \tau r_n \rangle}{\langle r_{n-1}, \tau r_{n-1} \rangle}.$$

This gives the standard CG formulation.

ALGORITHM 5.2.
The CG method

Initialize: Given $\mathscr{A} : V \to V^{\#}$ self-adjoint, bounded, and coercive, $b \in V^{\#}$, an initial
approximation $x_0 \in V$, the duality pairing $\langle \cdot, \cdot \rangle : V^{\#} \times V \to \mathbb{R}$, and the inner product
$(\cdot, \cdot)_V : V \times V \to \mathbb{R}$ determining the Riesz map τ,

$$\text{set } r_0 = b - \mathscr{A} x_0 \in V^{\#}, \quad p_0 = \tau r_0 \in V.$$

for $n = 1, 2, \ldots, n_{\max}$ **do**

$$\alpha_{n-1} = \frac{\langle r_{n-1}, \tau r_{n-1} \rangle}{\langle \mathscr{A} p_{n-1}, p_{n-1} \rangle} = \frac{(\tau r_{n-1}, \tau r_{n-1})_V}{(\tau \mathscr{A} p_{n-1}, p_{n-1})_V}$$

$$x_n = x_{n-1} + \alpha_{n-1} p_{n-1} \qquad \qquad \triangleright \text{ stop when the stopping criterion is satisfied}$$

$$r_n = r_{n-1} - \alpha_{n-1} \mathscr{A} p_{n-1}$$

$$\beta_n = \frac{\langle r_n, \tau r_n \rangle}{\langle r_{n-1}, \tau r_{n-1} \rangle} = \frac{(\tau r_n, \tau r_n)_V}{(\tau r_{n-1}, \tau r_{n-1})_V}$$

$$p_n = \tau r_n + \beta_n p_{n-1}$$

end for

By construction, the nth Krylov subspace can be equivalently written as

$$K_n(\tau \mathscr{A}, \tau r_0) = \text{span}\{\tau r_0, \tau \mathscr{A}(\tau r_0), \ldots, (\tau \mathscr{A})^{n-1}(\tau r_0)\}$$
$$= \text{span}\{p_0, p_1, \ldots, p_{n-1}\}$$
$$= \text{span}\{\tau r_0, \tau r_1, \ldots, \tau r_{n-1}\}$$

with $x_n \in x_0 + K_n(\tau \mathscr{A}, \tau r_0) \subset V$ and

$$r_n = b - \mathscr{A} x_n \in \text{span}\{r_0, \mathscr{A} \tau r_0, \ldots, (\mathscr{A} \tau)^{n-1} r_0\} \subset V^{\#}.$$

Remark 4. Krylov subspace methods in Hilbert spaces minimizing residuals can be easily
formulated using the relationship to the Lanczos or Arnoldi orhogonalization processes;
see [92]. For CG an analogous relationship is revealed in the next section, which also
demonstrates that the existence of short recurrences for the residuals and direction vectors
in Algorithm 5.2 is an immediate consequence of the link with the associated orthonormal
polynomials.

5.2 ▪ Analytic moment matching model reduction description of CG

We return to the Galerkin orthogonality condition (5.7). In what follows we will heavily use the ideas and results presented in the beautiful monograph by Vorobyev [192], originally published (in Russian) in 1958 and which unfortunately remains almost unknown in the mathematical community (in contrast to the attention paid to it in the physics literature; see [124, Section 3.7]).

Consider the first n elements of the Krylov sequence $\tau r_0, \tau\mathscr{A}(\tau r_0), \ldots, (\tau\mathscr{A})^{n-1}\tau r_0$ generating $K_n(\tau\mathscr{A}, \tau r_0)$. Let E_n be the orthogonal projection with respect to the inner product $(\cdot, \cdot)_V$ onto the nth Krylov subspace $K_n(\tau\mathscr{A}, \tau r_0) \subset V$. Then we can construct the orthogonally restricted operator[12]

$$\tau\mathscr{A}_n : K_n \to K_n$$

by formulating the following equalities:

$$
\begin{aligned}
\tau\mathscr{A}_n(\tau r_0) &:= \tau\mathscr{A}(\tau r_0), \\
(\tau\mathscr{A}_n)^2 \tau r_0 = \tau\mathscr{A}_n(\tau\mathscr{A}(\tau r_0)) &:= (\tau\mathscr{A})^2 \tau r_0, \\
&\;\;\vdots \\
(\tau\mathscr{A}_n)^{n-1} \tau r_0 = \tau\mathscr{A}_n((\tau\mathscr{A})^{n-2}\tau r_0) &:= (\tau\mathscr{A})^{n-1}\tau r_0, \\
(\tau\mathscr{A}_n)^n \tau r_0 = \tau\mathscr{A}_n((\tau\mathscr{A})^{n-1}\tau r_0) &:= E_n(\tau\mathscr{A})^n \tau r_0.
\end{aligned}
\tag{5.19}
$$

We will assume (for simplicity of notation and with no loss of generality) that K_n has dimension n. Then $\tau\mathscr{A}_n$ is uniquely determined by (5.19). Since it can be expressed as

$$\tau\mathscr{A}_n = E_n(\tau\mathscr{A})E_n : V \to V, \tag{5.20}$$

$\tau\mathscr{A}_n$ represents a self-adjoint linear operator in V which is bounded and coercive. The Galerkin orthogonality (5.7) gives (using $x_n = x_0 + z_n$, $z_n \in K_n$)

$$0 = E_n \tau r_n = E_n \tau r_0 - E_n \tau\mathscr{A} z_n = \tau r_0 - E_n \tau\mathscr{A} E_n z_n.$$

Consequently, x_n is determined by solving the n-dimensional linear system

$$\tau\mathscr{A}_n z_n = \tau r_0, \quad x_n = x_0 + z_n, \quad z_n \in K_n(\tau\mathscr{A}, \tau r_0), \tag{5.21}$$

which represents another formulation of CG.

A natural consequence of (5.19) is that the n-dimensional approximation $\tau\mathscr{A}_n$ to $\tau\mathscr{A}$ matches the first $2n$ moments

$$((\tau\mathscr{A}_n)^\ell \tau r_0, \tau r_0)_V = ((\tau\mathscr{A})^\ell \tau r_0, \tau r_0)_V, \quad \ell = 0, 1, \ldots, 2n - 1. \tag{5.22}$$

Indeed, for $\ell = 0, 1, \ldots, n-1$ this immediately follows from the first $n-1$ rows in (5.19). The last equation in (5.19) gives

$$E_n\{(\tau\mathscr{A}_n)^n \tau r_0 - (\tau\mathscr{A})^n \tau r_0\} = 0, \quad \text{i.e.,} \quad \{(\tau\mathscr{A}_n)^n \tau r_0 - (\tau\mathscr{A})^n \tau r_0\} \perp K_n.$$

---

[12]It should be understood that since K_n depends on τr_0, the orthogonally restricted operator $\tau\mathscr{A}_n$ is constructed for this particular τr_0 and does not represent in any way a general n-dimensional approximation of $\tau\mathscr{A}$.

Considering with the term $(\tau\mathscr{A}_n)^n\tau r_0$ the orthogonality with respect to the basis τr_0, $\ldots$, $(\tau\mathscr{A}_n)^{n-1}\tau r_0$ and with the term $(\tau\mathscr{A})^n\tau r_0$ the orthogonality with respect to the same basis but relabeled as $\tau r_0,\ldots,(\tau\mathscr{A})^{n-1}\tau r_0$, the rest of the statement is proved for $\ell = n,\ldots,2n-1$. We call (5.20)–(5.22) the *Vorobyev moment problem formulation of CG*.

It is worth mentioning that Luenberger refers to the book [192] of Vorobyev in his book published in 1969 [127]. In his description of CG he uses the term "orthogonalization of moments," but this in fact means orthogonalization of the Krylov sequence (see the last paragraph of Chapter 4). A more extensive interest in the method of Vorobyev then does not appear in mathematical literature until the works of Brezinski [32, 33] published in 1996 and 1997.

The finite-dimensional approximation $\tau\mathscr{A}_n$ of $\tau\mathscr{A}$ can be conveniently expressed via its *matrix* associated with a proper basis of $K_n(\tau\mathscr{A},\tau r_0)$. In particular, consider a sequence $q_1,q_2,\ldots$ such that $q_1 = \tau r_0/\|\tau r_0\|_V$ and $q_1,\ldots,q_\ell$ form an orthonormal basis of $K_\ell(\tau\mathscr{A},\tau r_0)$, $\ell = 1,\ldots,n$:

$$(q_i,q_j)_V = \delta_{ij}, \quad i,j = 1,2,\ldots,n. \tag{5.23}$$

Such a sequence can be determined using the Gram–Schmidt orthogonalization process, which for the self-adjoint $\tau\mathscr{A}$ reduces to the three-term recurrence of the Lanczos process [117, 118]. (The same three-term recurrence has been known in European mathematics in the context of continued fractions and orthogonal polynomials for several centuries and appeared, e.g., in the works of Brouncker and Wallis, Euler, Chebyshev, Christoffel, and Stieltjes; see [179, 31, 70][13].)

Denoting symbolically $Q_n = (q_1,\ldots,q_n)$ a matrix composed of the columns $q_1,\ldots,q_n$ and using the symbolic notation $q_j^* q_i := (q_i,q_j)_V$, $q_i,q_j \in V$, we can write $E_n = Q_n Q_n^*$, with the symbol $*$ denoting the standard algebraic transpose of the matrix (vector), and

$$\tau\mathscr{A}_n = Q_n Q_n^* \tau\mathscr{A} Q_n Q_n^* = Q_n \mathbf{T}_n Q_n^*, \tag{5.24}$$

where

$$\mathbf{T}_n = \begin{pmatrix} \gamma_1 & \delta_2 & & & \\ \delta_2 & \ddots & \ddots & & \\ & \ddots & \ddots & \ddots & \\ & & \ddots & \ddots & \delta_n \\ & & & \delta_n & \gamma_n \end{pmatrix} \tag{5.25}$$

is the Jacobi matrix of orthogonalization coefficients determined by the Lanczos process (here and elsewhere $\mathbf{e}_\ell$ denotes the ℓth vector of Euclidean basis)

$$\tau\mathscr{A}\, Q_n = Q_n \mathbf{T}_n + \delta_{n+1} q_{n+1} \mathbf{e}_n^*. \tag{5.26}$$

Since $q_1 = \tau r_0/\|\tau r_0\|$ and $(\tau\mathscr{A}_n)^\ell = Q_n \mathbf{T}_n^\ell Q_n^*$, $\ell = 0,1,\ldots$, the matching moments condition (5.22) is equivalently written as

$$\mathbf{e}_1^* \mathbf{T}_n^\ell \mathbf{e}_1 = q_1^*(\tau\mathscr{A})^\ell q_1, \quad \ell = 0,1,\ldots,2n-1. \tag{5.27}$$

Using (5.21) and (5.24), CG can thus be formulated in a *matrix form* as

$$\mathbf{T}_n \mathbf{y}_n = \|\tau r_0\|_V \mathbf{e}_1, \quad x_n = x_0 + Q_n \mathbf{y}_n, \tag{5.28}$$

---

[13]For a very rich context and further references we suggest reading the historical notes (in particular Sections 2.5.7 and 3.3.6 on p. 64 and p. 107) presented in the monograph [124].

where the (tridiagonal) matrix $\mathbf{T}_n$ is uniquely determined by the matching moments property (5.27). Consequently, CG can be seen as the *model reduction* which determines the nth approximate solution by solving the reduced algebraically formulated model (5.28) with $\mathbf{T}_n$ satisfying the first $2n$ *operator moments* (5.27).

The presented CG formulation (5.27)–(5.28) uses matrix representation of the reduced model. It is worth pointing out that even in the infinite-dimensional Hilbert space setting, the first n steps of CG are *always* given by the $n \times n$ matrix representation (5.28). In order to determine this matrix representation, the Lanczos process (5.26) in V is needed; see also Remark 2 at the beginning of this chapter.

In order to link CG with classical results in analysis and approximation theory (see, in particular, the fascinating paper of Stieltjes from 1894 [179]), we conveniently use spectral decompositions of $\tau\mathscr{A}$ and of $\mathbf{T}_n$. Since $\tau\mathscr{A}$ is self-adjoint and bounded, its spectral decomposition is written using the Riemann–Stieltjes integral as (see, e.g., [192, Chapter III, Section 7], [189], [190, Chapter II, Section 7], [124, Remark 3.5.1, Sections 4.9, 3.3.5 and 3.4.3], [131, 132], and the literature given there)

$$\tau\mathscr{A} = \int_{\lambda_L}^{\lambda_U} \lambda \, \mathrm{d}\mathscr{E}_\lambda, \tag{5.29}$$

where $0 < \lambda_L < \lambda_U$ are chosen arbitrarily such that the open interval (λ_L, λ_U) contains the spectrum of $\tau\mathscr{A}$ (which consist, in general, of the eigenvalues of $\tau\mathscr{A}$ and the continuous part of the spectrum). The spectral function $\mathscr{E}_\lambda$ of $\tau\mathscr{A}$ represents a family of orthogonal projections which is

(1) nondecreasing, i.e., if $\mu > \nu$, then the subspace onto which $\mathscr{E}_\mu$ projects contains the subspace into which $\mathscr{E}_\nu$ projects;

(2) $\mathscr{E}_{\lambda_L} = 0$, $\mathscr{E}_{\lambda_U} = I$;

(3) $\mathscr{E}_\lambda$ is right continuous, i.e., $\lim_{\lambda' \to \lambda_+} \mathscr{E}_{\lambda'} = \mathscr{E}_\lambda$.

The values of λ where $\mathscr{E}_\lambda$ increases by jumps represent the eigenvalues of $\tau\mathscr{A}$, with the eigenvectors satisfying

$$\tau\mathscr{A} z = \lambda z, \quad z \in V.$$

The points where $\mathscr{E}_\lambda$ increases continuously constitute a continuous part of the spectrum of $\tau\mathscr{A}$. For the (finite) Jacobi matrix $\mathbf{T}_n$ we can analogously write (by replacing the integral by the sum)

$$\mathbf{T}_n = \sum_{j=1}^{n} \theta_j^{(n)} \mathbf{s}_j^{(n)} (\mathbf{s}_j^{(n)})^*, \quad \lambda_L < \theta_1^{(n)} < \theta_2^{(n)} < \cdots < \theta_n^{(n)} < \lambda_U, \tag{5.30}$$

where $\mathbf{s}_1^{(n)}, \mathbf{s}_2^{(n)}, \ldots, \mathbf{s}_n^{(n)}$ are the associated orthonormal eigenvectors of the matrix $\mathbf{T}_n$, i.e., $(\mathbf{s}_i^{(n)})^* \mathbf{s}_j^{(n)} = \delta_{ij}$, $i,j = 1,\ldots,n$. Using these decompositions, the operator moment problem (5.27) is equivalently written as

$$\sum_{j=1}^{n} \{\theta_j^{(n)}\}^\ell \, \omega_j^{(n)} = \int_{\lambda_L}^{\lambda_U} \lambda^\ell \, \mathrm{d}\omega(\lambda), \quad \ell = 0, 1, \ldots, 2n - 1. \tag{5.31}$$

Here $\omega_j^{(n)} = ((\mathbf{s}_j^{(n)})^* \mathbf{e}_1)^2$, $\mathrm{d}\omega(\lambda) = q_1^* \, \mathrm{d}\mathscr{E}_\lambda \, q_1$, and $\omega(\lambda)$ represents the Riemann–Stieltjes spectral distribution function associated with $\tau\mathscr{A}$ and q_1 which is nondecreasing and right-continuous on the interval $[\lambda_L, \lambda_U]$, and $\omega(\lambda_L) = 0$, $\omega(\lambda_U) = 1$.

We see that CG can be seen as a method for computing the sequence of distribution functions $\omega^{(1)}, \omega^{(2)}, \ldots$, where $\omega^{(n)}$ is determined by the n points of increase $\theta_j^{(n)}$ and the values of the associated jumps $\omega_j^{(n)}$, $j = 1, \ldots, n$.

Since (5.31) holds for $\ell = 0, 1, \ldots, 2n - 1$, any linear combination of the monomials $1, \lambda, \ldots, \lambda^{2n-1}$ can be integrated exactly with respect to the distribution function $\omega(\lambda)$ simply by evaluating the value of this linear combination at the n district nodes $\theta_1^{(n)}, \ldots, \theta_n^{(n)}$ and then summing up with the weights $\omega_1^{(n)}, \ldots, \omega_n^{(n)}$, i.e., by summing up the left part of (5.31) multiplied by the polynomial coefficients. Summarizing, the distribution function $\omega^{(n)}(\lambda)$ approximates $\omega(\lambda)$ *in the sense of the nth Gauss–Christoffel quadrature*; see [69, 99, 44].

The relationship can be summarized by the following scheme (see also [124, Figure 3.2, p. 139]), where $f(\lambda)$ stands for a Riemann–Stieltjes integrable function:

$$\mathscr{A}, q_1 = \tau r_0 / \|\tau r_0\|_V \quad \longleftrightarrow \quad \omega(\lambda), \quad \int_{\lambda_L}^{\lambda_U} f(\lambda)\, d\omega(\lambda)$$

$$\downarrow \qquad\qquad\qquad\qquad\qquad\qquad \downarrow$$

$$\text{Lanczos/CG} \qquad\qquad \text{Gauss–Christoffel quadrature}$$

$$\downarrow \qquad\qquad\qquad\qquad\qquad\qquad \downarrow$$

$$\mathbf{T}_n, \mathbf{e}_1 \quad \longleftrightarrow \quad \omega^{(n)}(\lambda), \quad \sum_{j=1}^{n} \omega_j^{(n)} f(\theta_j^{(n)})$$

Here $\omega(\lambda)$ does not determine the spectral function $\mathscr{E}_\lambda$, i.e., it does not determine the operator $\mathscr{A}$ and the function q_1, but the decomposition $q_1^* \, d\mathscr{E}_\lambda q_1$ which is sufficient for determining the Jacobi matrices $\mathbf{T}_n$, $n = 1, 2, \ldots$. This is immediately clear from the fact that $\omega^{(n)}(\lambda)$ uniquely determines $\mathbf{T}_n$; for details see the summary in [124, Section 3.4.1]. If we take $f(\lambda) = \lambda^{-1}$, then the relationship between the CG iterations and the Gauss–Christoffel quadrature represents the following isometry:

$$\int_{\lambda_L}^{\lambda_U} \lambda^{-1} \, d\omega(\lambda) = \sum_{j=1}^{n} \omega_j^{(n)} \{\theta_j^{(n)}\}^{-1} + \frac{\|x - x_n\|_a^2}{\|\tau r_0\|_V^2}, \tag{5.32}$$

i.e., the error of the n-node Gauss–Christoffel quadrature is nothing but the squared $\|\cdot\|_a$ norm of the nth CG error normalized by $\|\tau r_0\|_V$. Indeed (see also Theorem 3.5.2 in [124, Section 3.5]), noticing that

$$\|x - x_0\|_a^2 = \langle \mathscr{A}(x - x_0), x - x_0 \rangle = (\tau \mathscr{A}(x - x_0), x - x_0)_V$$

$$= ((\tau \mathscr{A})^{-1} \tau r_0, \tau r_0)_V = \|\tau r_0\|_V^2 \int_{\lambda_L}^{\lambda_U} \lambda^{-1} \, d\omega(\lambda),$$

we observe

$$
\begin{aligned}
\frac{\|x-x_n\|_a^2}{\|\tau r_0\|_V^2}
&= \frac{\|x-x_0-Q_n\mathbf{y}_n\|_a^2}{\|\tau r_0\|_V^2}
= \frac{\|x-x_0\|_a^2}{\|\tau r_0\|_V^2}
- \frac{2(x-x_0,Q_n\mathbf{y}_n)_a-\|Q_n\mathbf{y}_n\|_a^2}{\|\tau r_0\|_V^2} \\[2mm]
&= \int_{\lambda_L}^{\lambda_U} \lambda^{-1}\,\mathrm{d}\omega(\lambda)
- \frac{2\langle \mathscr{A}(x-x_0),Q_n\mathbf{y}_n\rangle-\langle \mathscr{A}Q_n\mathbf{y}_n,Q_n\mathbf{y}_n\rangle}{\|\tau r_0\|_V^2} \\[2mm]
&= \int_{\lambda_L}^{\lambda_U} \lambda^{-1}\,\mathrm{d}\omega(\lambda)
- \frac{2(\tau\mathscr{A}(x-x_0),Q_n\mathbf{y}_n)_V-(\tau\mathscr{A}Q_n\mathbf{y}_n,Q_n\mathbf{y}_n)_V}{\|\tau r_0\|_V^2} \\[2mm]
&= \int_{\lambda_L}^{\lambda_U} \lambda^{-1}\,\mathrm{d}\omega(\lambda)
- \frac{2(\|\tau r_0\|_V Q_n\mathbf{e}_1,Q_n\mathbf{y}_n)_V-(Q_n^*\tau\mathscr{A}Q_n\mathbf{y}_n,\mathbf{y}_n)_V}{\|\tau r_0\|_V^2} \\[2mm]
&= \int_{\lambda_L}^{\lambda_U} \lambda^{-1}\,\mathrm{d}\omega(\lambda) \\[2mm]
&\quad - \frac{2(\|\tau r_0\|_V\mathbf{e}_1,\|\tau r_0\|_V \mathbf{T}_n^{-1}\mathbf{e}_1)_V-(\|\tau r_0\|_V\mathbf{e}_1,\|\tau r_0\|_V \mathbf{T}_n^{-1}\mathbf{e}_1)_V}{\|\tau r_0\|_V^2} \\[2mm]
&= \int_{\lambda_L}^{\lambda_U} \lambda^{-1}\,\mathrm{d}\omega(\lambda) - \mathbf{e}_1^*\mathbf{T}_n^{-1}\mathbf{e}_1
= \int_{\lambda_L}^{\lambda_U} \lambda^{-1}\,\mathrm{d}\omega(\lambda) - \sum_{j=1}^{n}\omega_j^{(n)}\{\theta_j^{(n)}\}^{-1},
\end{aligned}
$$

and (5.32) follows.

The outlined one-to-one connection of CG with the Gauss–Christoffel quadrature automatically brings in the whole richness of the theory of orthogonal polynomials, continued fractions, etc. mentioned above, which can then provide tools for the convergence analysis of CG. The isometry described above has a simple yet fundamental consequence which precedes the detailed discussion in Chapter 11 (see also Section 5.1):

> As the error in the Gauss–Christoffel quadrature cannot be expressed in terms of the operator condition number $\varkappa(\mathscr{A})$, no bound based solely on $\varkappa(\mathscr{A})$ without taking into account further information about the distribution function $\omega(\lambda)$ is generally descriptive in the evaluation of the rate of convergence of CG.

Further details on the outlined relationship can be found, e.g., in [192, in particular Chapter 5] and [124, in particular Chapters 3 and 5] and in the literature referred to there.

As mentioned in Section 5.1, an a priori quantification of the computational costs of CG computations is in general very difficult. A bound based on the operator condition number can be very pessimistic and may not allow for a realistic evaluation of the computational cost. The link with the Gauss–Christoffel quadrature, on the other hand, can qualitatively predict fast or slow convergence for problems with some particular spectral distribution functions.

Up to now we have considered mathematical properties of CG. Computational properties concerning the effects of rounding errors and incorporation of stopping criteria, which are fundamental for any practical use of CG, will be briefly addressed in Chapter 12.

Chapter 6

Finite-Dimensional Hilbert Spaces and the Matrix Formulation of the Conjugate Gradient Method

In the previous chapter the dimension of the Hilbert space V is, in general, infinite. Convergence of the finite-dimensional approximations $\mathcal{A}_n$ and x_n (which were not studied in the text above) can be associated with the convergence of the corresponding distribution functions $\omega^{(n)}(\lambda)$, $n = 1, 2, \ldots$, to $\omega(\lambda)$ in the sense of the Gauss–Christoffel quadrature; see [192]. For a finite-dimensional Hilbert space, for later convenience denoted below as V_h, CG can be described fully in terms of vectors and matrices.

Let the Hilbert space V_h be of dimension $N < \infty$. Consider its arbitrary *fixed* basis Φ written symbolically as $\Phi = (\phi_1, \ldots, \phi_N)$. Then each element $u \in V_h$ can be represented in $\mathbb{R}^N$ by its coordinates $\mathbf{u}$ in the basis Φ, symbolically written as

$$u = \Phi\mathbf{u},$$

where $\mathbf{u} \in \mathbb{R}^N$ is a column vector (recall our convention mentioned above that all algebraic vectors will always be column vectors). Consider further the canonical dual basis $\Phi^\# = (\phi_1^\#, \ldots, \phi_N^\#)$ of $V_h^\#$ associated with Φ, i.e.,

$$\langle \phi_i^\#, \phi_j \rangle = \delta_{ij}, \quad i, j = 1, \ldots, N.$$

Then (recalling that $\mathbf{z}^*$ means the transposition of the vector $\mathbf{z}$)

$$\mathbf{u} = (\langle \phi_1^\#, u \rangle, \ldots, \langle \phi_N^\#, u \rangle)^*.$$

The duality pairing $\langle \cdot, \cdot \rangle$, the inner product $(\cdot, \cdot)_V$, the bilinear form $a(\cdot, \cdot)$, the operator $\mathcal{A}$, and the Riesz map τ are represented in $\mathbb{R}^N$ as follows. For any $f = \Phi^\#\mathbf{f} \in V_h^\#$ and $v = \Phi\mathbf{v} \in V_h$ we simply get

$$\langle f, v \rangle = \langle \Phi^\#\mathbf{f}, \Phi\mathbf{v} \rangle = \mathbf{v}^* \left(\langle \phi_i^\#, \phi_j \rangle \right)_{i,j=1,\ldots,N} \mathbf{f} = \mathbf{v}^*\mathbf{f}. \tag{6.1}$$

Therefore the duality pairing $\langle \cdot, \cdot \rangle : V_h^\# \times V_h \to \mathbb{R}$ is represented by the standard Euclidean inner product in $\mathbb{R}^N$ (here we identify in the standard way $(\mathbb{R}^N)^\#$ with $\mathbb{R}^N$). Consider now the inner product $(\cdot, \cdot)_{V_h}$ in V_h.[14] For any $u = \Phi\mathbf{u}$, $v = \Phi\mathbf{v} \in V_h$,

$$(u, v)_{V_h} = (\Phi\mathbf{u}, \Phi\mathbf{v})_{V_h} = \mathbf{v}^*\mathbf{M}\mathbf{u}, \quad \mathbf{M} := (\mathbf{M}_{ij}) := \left((\phi_j, \phi_i)_{V_h} \right)_{i,j=1,\ldots,N}. \tag{6.2}$$

---

[14] When discussing the relationship between preconditioning and transformation of the discretization bases in Chapter 8, it will be convenient to consider inner products on V_h different from the inner product $(\cdot, \cdot)_V$ on V used in the derivation of CG in Chapter 5. Therefore we use here the notation $(\cdot, \cdot)_{V_h}$ instead of $(\cdot, \cdot)_V$.

Here $\mathbf{M}$ is the Gram matrix of the basis Φ and therefore is symmetric and positive definite. For the bilinear form $a(\cdot,\cdot)$ and the associated operator $\mathscr{A}$,

$$\mathscr{A}\phi_k = \sum_{\ell=1}^{N} v_\ell\,\phi_\ell^{\#}, \quad v_\ell = \langle \mathscr{A}\phi_k, \phi_\ell \rangle = a(\phi_k, \phi_\ell).$$

Denoting

$$\mathbf{A} := (\mathbf{A}_{ij}) := \left(\langle \mathscr{A}\phi_j, \phi_i \rangle \right)_{i,j=1,\ldots,N} = \left(a(\phi_j, \phi_i) \right)_{i,j=1,\ldots,N}, \tag{6.3}$$

we immediately get for $\mathscr{A}\Phi : \mathbb{R}^N \to V_h^{\#}$

$$\mathscr{A}\Phi := (\mathscr{A}\phi_1, \ldots, \mathscr{A}\phi_N) = \Phi^{\#}\mathbf{A}; \tag{6.4}$$

i.e., the coordinates of the image $\mathscr{A}\phi_k$ in the basis $\Phi^{\#}$ are given by the kth column of the matrix $\mathbf{A}$. For any $\mathbf{z} \in \mathbb{R}^N$ we therefore have $\mathbf{A}\mathbf{z} \in (\mathbb{R}^N)^{\#}$ as coordinates of $\mathscr{A}\Phi\mathbf{z}$ in the dual basis $\Phi^{\#}$, which conforms with the representation of the duality pairing in $\mathbb{R}^N$. Indeed, for any $v \in V_h$

$$\langle \mathscr{A}\Phi\mathbf{z}, v \rangle = \langle \Phi^{\#}\mathbf{A}\mathbf{z}, \Phi\mathbf{v} \rangle = \mathbf{v}^{*}\mathbf{A}\mathbf{z}.$$

Since $\mathscr{A}$ is self-adjoint, bounded, and α-coercive, this also immediately gives that the matrix $\mathbf{A}$ is symmetric positive definite with

$$\mathbf{v}^{*}\mathbf{A}\mathbf{v} \geq \alpha\|v\|_{V}^{2}, \qquad v = \Phi\mathbf{v},$$

for any $\mathbf{v} \in \mathbb{R}^N$; see (3.16).

In order to get the representation for the Riesz map τ, consider an arbitrary $f \in V_h^{\#}$, $v \in V_h$, and recall the duality pairing representation (6.1):

$$\langle f, v \rangle = \langle \Phi^{\#}\mathbf{f}, \Phi\mathbf{v} \rangle = \mathbf{v}^{*}\mathbf{f}.$$

At the same time we have

$$\langle f, v \rangle = (\tau f, v)_{V_h} = (\tau\Phi^{\#}\mathbf{f}, \Phi\mathbf{v})_{V_h},$$

where $\tau\Phi^{\#} : \mathbb{R}^N \to V_h$ can be expressed using the basis Φ and the matrix

$$\mathbf{M}_{\tau} := (\Phi^{\#})^{*}(\tau\phi_1^{\#}, \ldots, \tau\phi_N^{\#}),$$

storing in its columns the coordinates of the functions $\tau\phi_\ell^{\#}$ in this basis, $\ell = 1,\ldots,N$, i.e.,

$$\tau\Phi^{\#} = \Phi\mathbf{M}_{\tau}.$$

Using this,

$$\mathbf{v}^{*}\mathbf{f} = \langle f, v \rangle = (\tau f, v)_{V_h} = (\tau\Phi^{\#}\mathbf{f}, \Phi\mathbf{v})_{V_h} = (\Phi\mathbf{M}_{\tau}\mathbf{f}, \Phi\mathbf{v})_{V_h} = \mathbf{v}^{*}\mathbf{M}\mathbf{M}_{\tau}\mathbf{f},$$

where we have used the inner product representation (6.2). Comparing the left and the right term, $\mathbf{M}_{\tau} = \mathbf{M}^{-1}$ is the matrix representation of the Riesz map τ, i.e.,

$$\tau\Phi^{\#} = \Phi\mathbf{M}^{-1}. \tag{6.5}$$

With these matrix and vector representations and rewriting (5.1) as

$$\mathscr{A}x = \mathscr{A}\Phi\mathbf{x} = \Phi^{\#}\mathbf{A}\mathbf{x} = b = \Phi^{\#}\mathbf{b},$$

we get the linear algebraic system

$$\mathbf{A}\mathbf{x} = \mathbf{b}, \quad \mathbf{A} \in \mathbb{R}^{N \times N}, \quad \mathbf{x} \in \mathbb{R}^N, \quad \mathbf{b} \in \mathbb{R}^N, \tag{6.6}$$

and we can reformulate Algorithm 5.2 as a finite-dimensional, purely algebraic, representation of CG for solving (6.6).

Taking $r_0 = b - \mathscr{A} x_0 = \Phi^{\#}(\mathbf{b} - \mathbf{A}\mathbf{x}_0) = \Phi^{\#}\mathbf{r}_0$, $x_0 = \Phi\mathbf{x}_0$, $p_0 = \tau r_0 = \tau\Phi^{\#}\mathbf{r}_0 = \Phi\mathbf{M}^{-1}\mathbf{r}_0 = \Phi\mathbf{p}_0$, where $\mathbf{r}_0 := \mathbf{b} - \mathbf{A}\mathbf{x}_0$ and $\mathbf{p}_0 := \mathbf{M}^{-1}\mathbf{r}_0$, we get

$$\langle r_{n-1}, \tau\, r_{n-1}\rangle = \langle \Phi^{\#}\mathbf{r}_{n-1}, \tau\Phi^{\#}\mathbf{r}_{n-1}\rangle = \langle \Phi^{\#}\mathbf{r}_{n-1}, \Phi\mathbf{M}^{-1}\mathbf{r}_{n-1}\rangle$$
$$= (\mathbf{M}^{-1}\mathbf{r}_{n-1})^{*}\mathbf{r}_{n-1},$$
$$\langle \mathscr{A} p_{n-1}, p_{n-1}\rangle = \langle \Phi^{\#}\mathbf{A}\mathbf{p}_{n-1}, \Phi\mathbf{p}_{n-1}\rangle = \mathbf{p}_{n-1}^{*}\mathbf{A}\mathbf{p}_{n-1},$$
$$x_n = x_{n-1} + \alpha_{n-1}p_{n-1} = \Phi(\mathbf{x}_{n-1} + \alpha_{n-1}\mathbf{p}_{n-1}) = \Phi\mathbf{x}_n,$$
$$r_n = r_{n-1} - \alpha_{n-1}\mathscr{A} p_{n-1} = \Phi^{\#}(\mathbf{r}_{n-1} - \alpha_{n-1}\mathbf{A}\mathbf{p}_{n-1}) = \Phi^{\#}\mathbf{r}_n,$$
$$p_n = \tau r_n + \beta_n p_{n-1} = \tau\Phi^{\#}\mathbf{r}_n + \beta_n\Phi\mathbf{p}_{n-1} = \Phi(\mathbf{M}^{-1}\mathbf{r}_n + \beta_n\mathbf{p}_{n-1}),$$

where we have used the coordinates of the elements of V_b and $V_b^{\#}$ in the corresponding bases Φ and $\Phi^{\#}$. Summarizing, CG for solving (6.6) is given as follows.

ALGORITHM 6.1.

CG for solving linear algebraic systems

Initialize: Given $\mathbf{A}\mathbf{x} = \mathbf{b}$, $\mathbf{A} \in \mathbb{R}^{N \times N}$, $\mathbf{b} \in \mathbb{R}^N$, $\mathbf{A}$ symmetric positive definite, $\mathbf{M}$ symmetric positive definite, and $\mathbf{x}_0 \in \mathbb{R}^N$ as above,

$$\text{set } \mathbf{r}_0 = \mathbf{b} - \mathbf{A}\mathbf{x}_0, \quad \text{solve } \mathbf{M}\mathbf{z}_0 = \mathbf{r}_0, \quad \text{set } \mathbf{p}_0 = \mathbf{z}_0.$$

for $n = 1, 2, \ldots, n_{\max}$ **do**

$$\alpha_{n-1} = \frac{\mathbf{z}_{n-1}^{*}\mathbf{r}_{n-1}}{\mathbf{p}_{n-1}^{*}\mathbf{A}\mathbf{p}_{n-1}}$$

$$\mathbf{x}_n = \mathbf{x}_{n-1} + \alpha_{n-1}\mathbf{p}_{n-1} \qquad\qquad \triangleright \text{ stop when the stopping criterion is satisfied}$$

$$\mathbf{r}_n = \mathbf{r}_{n-1} - \alpha_{n-1}\mathbf{A}\mathbf{p}_{n-1}$$

$$\text{solve } \mathbf{M}\mathbf{z}_n = \mathbf{r}_n$$

$$\beta_n = \frac{\mathbf{z}_n^{*}\mathbf{r}_n}{\mathbf{z}_{n-1}^{*}\mathbf{r}_{n-1}}$$

$$\mathbf{p}_n = \mathbf{z}_n + \beta_n\mathbf{p}_{n-1}$$

end for

Algorithm 6.1 represents a standard form of the *preconditioned* conjugate gradient method (PCG) for solving linear algebraic systems $\mathbf{A}\mathbf{x} = \mathbf{b}$ with symmetric positive definite $\mathbf{A}$ presented in literature. Recalling Remark 2 before Section 5.1, the setting gives PCG minimizing the energy norm of the error

$$\|\mathbf{x} - \mathbf{x}_n\|_{\mathbf{A}} := ((\mathbf{x} - \mathbf{x}_n)^{*}\mathbf{A}(\mathbf{x} - \mathbf{x}_n))^{\frac{1}{2}}$$

over the affine subspaces $\mathbf{x}_0 + K_n(\mathbf{M}^{-1}\mathbf{A}, \mathbf{M}^{-1}\mathbf{r}_0)$, $n = 1, 2, \ldots$, as a straightforward consequence of the choice of the duality pairing $\langle \cdot, \cdot \rangle$ and the inner product $(\cdot, \cdot)_{V_h}$.[15]

If the Hilbert space V is infinite-dimensional, then the computation with the operators from Chapter 5 is, in general, infeasible. Instead of the infeasible application of the infinite-dimensional CG to the infinite-dimensional problem (5.1), a feasible alternative is to find the *finite-dimensional approximation* of the problem (5.1) and then use Algorithm 6.1 as a tool for solving the associated *matrix problem* $\mathbf{Ax} = \mathbf{b}$. For infinite-dimensional Hilbert spaces V, a possible approach therefore consists of finding the finite-dimensional approximating subspace $V_h \subset V$ and of finding its basis Φ_h such that this choice allows to construct a sufficiently accurate numerical approximation to the solution x of (5.1) via solving *numerically* the resulting linear algebraic system $\mathbf{Ax} = \mathbf{b}$. Here the requirement is that this algebraic system can be numerically solved to an acceptable accuracy at an acceptable cost; see also the discussion in Chapter 4.

The state-of-the-art literature on using CG (as well as on using other Krylov subspace methods) proceeds in most cases (in solving BVP) in the following way. First the linear algebraic system $\mathbf{Ax} = \mathbf{b}$ is formed by some form of discretization independently of any algebraic method which is subsequently used for its numerical solution. Then the *standard unpreconditioned CG*, i.e., Algorithm 6.1 with $\mathbf{M} = \mathbf{I}$, is considered. Since in almost all cases this would result in a slow convergence of $\mathbf{x}_n$ to the algebraic solution $\mathbf{x}$, preconditioning of the algebraic problem is introduced to accelerate the convergence behavior. If the algebraic preconditioning is motivated by the operator preconditioning as described above, then this can be illustrated within our framework by the following scheme:

$$\{\mathscr{A}, b, \tau\} \to \{\mathscr{A}_h, b_h, \tau\} \to \{\mathbf{A}_h, \mathbf{b}_h, \mathbf{M}_h\} \to \text{Algorithm 6.1},$$

where h typically stands for a discretization parameter and $\mathbf{M}_h$ is linked with the choice of the inner product $(\cdot, \cdot)_{V_h}$, i.e., with the Riesz map τ.[16]

A common view, however, considers the bilinear form representation (3.5), forms the linear algebraic system (6.6) and looks for algebraic preconditioning which should be applied together with the *unpreconditioned* CG, i.e., it proceeds as

$$a(\cdot, \cdot), \langle b, \cdot \rangle \to \mathbf{A}_h, \mathbf{b}_h \to \text{construction of an algebraic preconditioner} \to \text{Algorithm 6.1}.$$

In this way, preconditioning is considered without being part of the discretization of the problem. The whole computational process for solving the original infinite-dimensional problem (5.1), construction of efficient algorithms, and their analysis is then broken into (more or less artificially separated) pieces.

Operator preconditioning deals with operators and PDEs, and algebraic preconditioning deals with matrices. Discretization can be considered a tool for getting from the infinite-dimensional operator setting to the finite-dimensional matrix algebraic setting. (Efficient algebraic preconditioning associated with the infinite-dimensional operator preconditioning can be derived in some cases in a purely algebraic way.)

Following the practice in solving challenging problems, preconditioning and discretization should be linked together as much as possible. In particular, as shown below, preconditioning can be interpreted within the framework of Chapters 5 and 6 as *transformation of the discretization basis* accompanied by the associated change of the inner product in V_h. The ideas linking preconditioning with a change of the discretization basis are

---

[15] Here and in the next two chapters it is worth pointing out to the analogous development presented in a bit different context by Faber, Manteuffel, and Parter in the paper quoted several times above; see [63, pp. 113, 118, 143–145].

[16] Here $\mathscr{A}_h$ should not be confused with $\mathscr{A}_n$ determined by the Vorobyev moment problem; see Section 5.2.

known, e.g., from the hierarchical basis preconditioning [199, 202, 200, 201] or the so-called projector preconditioning of the FETI-DP algorithms [100], with more examples and references given in Chapter 8. The next two chapters will develop them in a more general context.

Chapter 7

Comments on the Galerkin Discretization

We will first briefly summarize the discretization step from V to V_h and from $\mathscr{A}x = b$ to $\mathbf{Ax} = \mathbf{b}$. Consider an N-dimensional subspace $V_h \subset V$ with the duality pairing, the inner product, and the Riesz map τ. Let $\Phi_h = (\phi_1^{(h)}, \ldots, \phi_N^{(h)})$ be the basis of V_h, $\Phi_h^{\#} = (\phi_1^{(h)\#}, \ldots, \phi_N^{(h)\#})$ the associated canonical basis of its dual $V_h^{\#}$. Here the index h is used for later convenience. In relation to the finite element method (FEM) it characterizes in some way the size of the mesh elements. In this chapter the index h has no such meaning and, where appropriate (such as in the notation of the basis functions), it will be skipped for simplicity of notation.

The standard way of discretization of the weak BVP formulation (3.5) is to restrict the space of the approximations u_h (to the solution u) to V_h with the test functions v taken from the same subspace, i.e., to look for $u_h \in V_h$ such that

$$a(u_h, v) = \langle b, v \rangle \quad \text{for all} \quad v \in V_h. \tag{7.1}$$

Analogously to Section 3.2 (see (3.6)), the bilinear form $a(\cdot, \cdot) : V_h \times V_h \to \mathbb{R}$ defines the operator $\mathscr{A}_h : V_h \to V_h^{\#}$ such that

$$\langle \mathscr{A}_h u, v \rangle = a(u, v) \quad \text{for all} \quad u, v \in V_h, \tag{7.2}$$

which gives the following equivalent form of (7.1):

$$\langle \mathscr{A}_h u_h, v \rangle = \langle b, v \rangle \quad \text{for all} \quad v \in V_h. \tag{7.3}$$

By restricting the functional b to $V_h^{\#}$, i.e., by defining $b_h \in V_h^{\#}$: $\langle b_h, v \rangle := \langle b, v \rangle$ for all $v \in V_h$, (7.3) can be written in the operator form

$$\mathscr{A}_h u_h = b_h, \quad u_h \in V_h, \quad b_h \in V_h^{\#}, \quad \mathscr{A}_h : V_h \to V_h^{\#}. \tag{7.4}$$

Using the bases Φ and $\Phi^{\#}$, the equivalent formulations (7.1)–(7.4) of the problem on the finite-dimensional space V_h can now be *numerically* solved by application of CG to the linear algebraic system $\mathbf{Ax} = \mathbf{b}$ given by (6.6); see Chapter 6. For completeness it is natural to ask whether the operator $\mathscr{A} : V \to V^{\#}$ restricted to V_h determines via (6.3) the same matrix as the operator $\mathscr{A}_h : V_h \to V_h^{\#}$ determined from (7.1) and (7.2). This is immediately confirmed by construction,

$$\langle \mathscr{A} u, v \rangle = a(u, v) = \langle \mathscr{A}_h u, v \rangle \quad \text{for all} \quad u, v \in V_h. \tag{7.5}$$

The discretization described above can also be interpreted as the *enforcing orthogonality of the operator residual* to the discretization subspace V_h with respect to the duality pairing $\langle \cdot, \cdot \rangle$. Indeed, subtracting $a(u_h, v) = \langle b_h, v \rangle$ and $a(u, v) = \langle b_h, v \rangle$, where $v \in V_h$, we get

$$a(u_h - u, v) = 0 \quad \text{for all} \quad v \in V_h, \tag{7.6}$$

i.e,

$$\langle \mathscr{A} u_h - b, v \rangle = 0 \quad \text{for all} \quad v \in V_h. \tag{7.7}$$

Also, obviously,

$$\langle \mathscr{A} u_h - \mathscr{A} u, v \rangle = 0 \quad \text{for all} \quad v \in V_h, \tag{7.8}$$

which we frequently use below in the sense that we replace b by $\mathscr{A} u$ and vice versa. That means that the approximate solution $u_h \in V_h$ satisfies the Galerkin orthogonality (7.6)–(7.7). In other words, the discretized approximation u_h gives the residual $b - \mathscr{A} u_h \in V_h^{\#}$ which is orthogonal to the subspace V_h with respect to the duality pairing $\langle \cdot, \cdot \rangle$; see the Galerkin orthogonality in CG represented by (5.7).

All this is obviously present in any standard textbook on variational methods or FEM. It is included here for two reasons. First, we wish to emphasize the *independence of the abstract subspace setting* in the Galerkin discretization presented above of any mesh-motivated description typically presented in texts on FEM. We will return to this point in Chapter 10. Second, it should be pointed out that the outlined approach has to deal with two sources of error. *Discretization errors* account for the inaccuracies in approximating the infinite-dimensional (linear) problem in V by the finite-dimensional problem in V_h. *Computational errors* account for inaccuracies in solving the linear algebraic system; they are caused by rounding errors and by possible truncation errors in iterative computations.

The Galerkin orthogonality (7.6)–(7.7) is expressed in terms of the functional residual. Any professional implementation of direct algebraic solvers provides residuals proportional to machine precision. It may therefore seem suggestive that algebraic residuals conveniently measure computational errors. This is, however, one of the most common myths in scientific computing, which negatively affects analysis and comparison of algorithms as well as computations. One can even find beliefs that algebraic iterative methods for solving systems of linear equations can compute approximate solutions with an arbitrary accuracy, where the (incorrect!) argument is built upon the size of the iteratively updated residuals, and beliefs that direct methods producing algebraic residuals proportional in size to machine precision can in practical computations be considered accurate enough so that the associated algebraic errors need not be considered. We will return to the question of algebraic errors in Chapter 12.

Chapter 8

Preconditioning of the Algebraic System as Transformation of the Discretization Basis

Consider the Galerkin discretization (7.4) of (5.1). Chapter 6 describes the algebraic matrix-vector representation $\mathbf{Ax} = \mathbf{b}$ of the finite-dimensional problem $\mathcal{A}_h x_h = b_h$ and of CG in $\mathbb{R}^N$, where $N = \dim V_h$. Summarizing this representation, we have (considering $V_h \subset V, (\cdot,\cdot)_{V_h} = (\cdot,\cdot)_V$)

$$\langle b,v \rangle \to \mathbf{v}^* \mathbf{b},$$

$$(u,v)_V \to \mathbf{v}^* \mathbf{Mu}, \quad (\mathbf{M}_{ij}) = \left((\phi_j,\phi_i)_V \right)_{i,j=1,\ldots,N},$$

$$x_h \to \mathbf{x},$$

$$r_n \to \mathbf{r}_n,$$

$$p_n \to \mathbf{p}_n,$$

$$x_n \to \mathbf{x}_n,$$

$$\tau f \to \mathbf{M}^{-1}\mathbf{f},$$

$$\mathcal{A}_h u = \mathcal{A} u \to \mathbf{Au}, \quad (\mathbf{A}_{ij}) = \left(a(\phi_j,\phi_i) \right)_{i,j=1,\ldots,N} = \left(\langle \mathcal{A}\phi_j,\phi_i \rangle \right)_{i,j=1,\ldots,N},$$

where the last two relationships follow from the equalities (see Chapter 6 above; we drop the subscript h in denoting the bases and use simply Φ and $\Phi^\#$)

$$\tau f = \tau \Phi^\# \mathbf{f} = \Phi \mathbf{M}^{-1}\mathbf{f}, \tag{8.1}$$

$$\mathcal{A} u = \mathcal{A} \Phi \mathbf{u} = \Phi^\# \mathbf{Au}. \tag{8.2}$$

The algebraic CG method implemented as in Algorithm 6.1 (PCG) then solves the linear algebraic system $\mathbf{Ax} = \mathbf{b}$.

This is, however, not the way CG is derived and presented in the literature. (For exceptions see the references given above, in particular [92].) Instead, CG is presented among the methods solving linear algebraic systems which can possibly arise from the discretization of PDEs. But even in that case no particular attention is paid to the infinite-dimensional CG representation, its discretization, and preconditioning as an inherent part of the basic formulation of CG. The standard form of CG in the algebraic literature is as follows; see [96] and, e.g., [172, 83, 134, 11].

ALGORITHM 8.1.

The standard algebraic CG formulation

Initialize: Given $\mathbf{Ax} = \mathbf{b}$, $\mathbf{A} \in \mathbb{R}^{N \times N}$ symmetric positive definite, $\mathbf{b} \in \mathbb{R}^N$, $\mathbf{x}_0 \in \mathbb{R}^N$,

$$\text{set } \mathbf{r}_0 = \mathbf{b} - \mathbf{Ax}_0, \quad \text{set } \mathbf{p}_0 = \mathbf{r}_0.$$

for $n = 1, 2, \ldots, n_{\max}$ **do**

$$\alpha_{n-1} = \frac{\mathbf{r}_{n-1}^* \mathbf{r}_{n-1}}{\mathbf{p}_{n-1}^* \mathbf{A} \mathbf{p}_{n-1}}$$

$$\mathbf{x}_n = \mathbf{x}_{n-1} + \alpha_{n-1} \mathbf{p}_{n-1} \qquad\qquad \triangleright \text{ stop when the stopping criterion is satisfied}$$

$$\mathbf{r}_n = \mathbf{r}_{n-1} - \alpha_{n-1} \mathbf{A} \mathbf{p}_{n-1}$$

$$\beta_n = \frac{\mathbf{r}_n^* \mathbf{r}_n}{\mathbf{r}_{n-1}^* \mathbf{r}_{n-1}}$$

$$\mathbf{p}_n = \mathbf{r}_n + \beta_n \mathbf{p}_{n-1}$$

end for

Recall that Algorithm 8.1 is obtained from Algorithm 6.1 simply by setting $\mathbf{M} = \mathbf{I}$. In the derivation using the functional setting of Chapters 5 and 6 this means that the basis functions $\phi_1, \ldots, \phi_N$ are *orthonormal with respect to the inner product* $(\cdot, \cdot)_V$.

The situation that the discretization basis vectors are orthonormal with respect to the inner product $(\cdot, \cdot)_V$ is certainly uncommon. We will consider a transformation (i.e., orthogonalization) of the discretization basis such that this condition is satisfied. Since $\mathbf{M}$ is the Gram matrix of the basis Φ (see (6.2)), the orthogonalization coefficients are given by the columns of the upper triangular matrix $(\mathbf{L}^*)^{-1}$, where $\mathbf{M} = \mathbf{L}\mathbf{L}^*$ is the Cholesky decomposition of $\mathbf{M}$; see [79, Section 4] and [124, Section 3.6]. Indeed, with

$$\Phi_t = \Phi(\mathbf{L}^*)^{-1}$$

we have (using the symbolic notation and $\mathbf{M} = \Phi^*\Phi$)

$$\Phi_t^* \Phi_t = \mathbf{L}^{-1}\Phi^*\Phi(\mathbf{L}^*)^{-1} = \mathbf{L}^{-1}\mathbf{M}(\mathbf{L}^*)^{-1} = \mathbf{I}.$$

Using the discretization basis Φ_t instead of Φ will lead to changing of the associated canonical dual basis to

$$\Phi_t^\# = \Phi^\# \mathbf{L}.$$

Indeed, with $\Phi_t = (\phi_1^t, \ldots, \phi_N^t)$ and $\Phi_t^\# = (\phi_1^{\#t}, \ldots, \phi_N^{\#t})$,

$$\langle \phi_i^{\#t}, \phi_j^t \rangle = \langle \Phi^\# \mathbf{L} \mathbf{e}_i, \Phi(\mathbf{L}^*)^{-1} \mathbf{e}_j \rangle = ((\mathbf{L}^*)^{-1}\mathbf{e}_j)^* \mathbf{L} \mathbf{e}_i = (\mathbf{e}_j)^* \mathbf{e}_i = \delta_{ij}.$$

Then

$$b_h = \Phi^\# \mathbf{b} = \Phi_t^\# \mathbf{L}^{-1}\mathbf{b} = \Phi_t^\# \mathbf{b}^t, \qquad \mathbf{b}^t = \mathbf{L}^{-1}\mathbf{b},$$

$$v = \Phi \mathbf{v} = \Phi_t \mathbf{L}^* \mathbf{v} = \Phi_t \mathbf{v}^t, \qquad \mathbf{v}^t = \mathbf{L}^* \mathbf{v},$$

$$x_h = \Phi \mathbf{x} = \Phi_t \mathbf{x}^t, \qquad \mathbf{x}^t = \mathbf{L}^* \mathbf{x},$$

$$r_n = \Phi^\# \mathbf{r}_n = \Phi_t^\# \mathbf{r}_n^t, \qquad \mathbf{r}_n^t = \mathbf{L}^{-1}\mathbf{r}_n,$$

$$p_n = \Phi \mathbf{p}_n = \Phi_t \mathbf{p}_n^t, \qquad \mathbf{p}_n^t = \mathbf{L}^* \mathbf{p}_n,$$

$$x_n = \Phi \mathbf{x}_n = \Phi_t \mathbf{x}_n^t, \qquad \mathbf{x}_n^t = \mathbf{L}^* \mathbf{x}_n,$$

and

$$\tau f = \tau \Phi^{\#} \mathbf{f} = \tau \Phi_t^{\#} \mathbf{L}^{-1} \mathbf{f} = \tau \Phi_t^{\#} \mathbf{f}^t$$
$$= \Phi \mathbf{M}^{-1} \mathbf{f} = \Phi_t \mathbf{L}^* \mathbf{M}^{-1} \mathbf{L} \mathbf{f}^t = \Phi_t \mathbf{f}^t, \qquad \mathbf{M}_t = \mathbf{I},$$
$$\mathscr{A} w = \mathscr{A} \Phi \mathbf{w} = \mathscr{A} \Phi_t \mathbf{L}^* \mathbf{w} = \mathscr{A} \Phi_t \mathbf{w}^t$$
$$= \Phi^{\#} \mathbf{A} \mathbf{w} = \Phi_t^{\#} \mathbf{L}^{-1} \mathbf{A} (\mathbf{L}^*)^{-1} \mathbf{w}^t$$
$$= \Phi_t^{\#} \mathbf{A}_t \mathbf{w}^t, \qquad\qquad \mathbf{A}_t = \mathbf{L}^{-1} \mathbf{A} (\mathbf{L}^*)^{-1}.$$

Consequently, we obtain the matrix representation of the Algorithm 5.2 applied to the finite-dimensional problem $\mathscr{A}_h x_h = b_h$ with the transformed (orthonormal) discretization basis Φ_t as the standard unpreconditioned algebraic CG (Algorithm 8.1) applied to the *preconditioned system*

$$\mathbf{A}_t \mathbf{x}^t = \mathbf{b}^t, \qquad \text{i.e.,} \qquad (\mathbf{L}^{-1} \mathbf{A} (\mathbf{L}^*)^{-1})(\mathbf{L}^* \mathbf{x}) = \mathbf{L}^{-1} \mathbf{b}. \qquad (8.3)$$

This observation can be interpreted in the following way:

Algebraic preconditioning of the linear algebraic system associated with the operator preconditioning is equivalent to orthonormalization of the discretization basis in the given finite-dimensional Hilbert space. In the context of the numerical solution of PDEs this means that discretization and preconditioning are tightly linked together.

In other words, *if we know the orthonormal discretization basis Φ_t beforehand*, then using this basis will provide preconditioning, and application of the unpreconditioned CG to $\mathbf{A}_t \mathbf{x}^t = \mathbf{b}^t$ is equivalent to application of Algorithm 6.1 to $\mathbf{A} \mathbf{x} = \mathbf{b}$.

For any choice of the finite-dimensional subspace $V_h \subset V$, the Galerkin discretization (7.4), and for any choice of the operator preconditioning determined by the inner product $(\cdot, \cdot)$ in V, the condition number of the matrix $\mathbf{M}^{-1} \mathbf{A}$, or, equivalently, the condition number of the matrix $\mathbf{A}_t$, is bounded by the condition number of the operator $\mathscr{A}_h$ given in (7.4) and therefore by the condition number of the operator $\mathscr{A}$; see Chapter 4. Indeed, using the fact that the basis Φ_t is orthonormal and therefore for any vector $\mathbf{u} \in \mathbb{R}^N$ we have that $\|u\|_V = \|\Phi_t \mathbf{u}\|_V = \|\mathbf{u}\| := (\mathbf{u}^* \mathbf{u})^{1/2}$, $u = \Phi_t \mathbf{u}$, and

$$\varkappa(\mathbf{M}^{-1}\mathbf{A}) = \varkappa(\mathbf{A}_t) = \frac{\max_{\|\mathbf{u}\|=1} \mathbf{u}^* \mathbf{A}_t \mathbf{u}}{\min_{\|\mathbf{v}\|=1} \mathbf{v}^* \mathbf{A}_t \mathbf{v}}$$
$$= \frac{\max_{\|\mathbf{u}\|=1} \mathbf{u}^* ((\mathscr{A} \phi_j^t, \phi_i^t))_{i,j=1,\ldots,N} \mathbf{u}}{\min_{\|\mathbf{v}\|=1} \mathbf{v}^* ((\mathscr{A} \phi_j^t, \phi_i^t))_{i,j=1,\ldots,N} \mathbf{v}}$$
$$= \frac{\max_{u \in V_h;\|u\|_V=1} \langle \mathscr{A} u, u \rangle}{\min_{v \in V_h;\|v\|_V=1} \langle \mathscr{A} v, v \rangle}$$
$$\leq \frac{\sup_{u,v \in V;\|u\|_V=\|v\|_V=1} |\langle \mathscr{A} u, v \rangle|}{\inf_{v \in V;\|v\|_V=1} \langle \mathscr{A} v, v \rangle}$$
$$= \varkappa(\mathscr{A});$$

for a more general statement (in a more general setting) we refer, e.g., to [98, Theorem 2.1 and relation (3.2)].

In derivations of algebraic preconditioning and in its incorporation into the whole solution process the link with discretization is typically not made and algebraic preconditioning is cut off from the discretization step. The linear algebraic system $\mathbf{Ax} = \mathbf{b}$ resulting from discretization of $\mathscr{A}x = b$, or, more precisely, from discretization of the bilinear form representation $a(x, v) = \langle b, v \rangle$ using some standard techniques such as the Galerkin finite element method, is further considered autonomously as a separate algebraic problem. If there is a need for preconditioning (which is almost always the case), then $\mathbf{Ax} = \mathbf{b}$ is replaced by

$$(\widehat{\mathbf{L}}^{-1}\mathbf{A}(\widehat{\mathbf{L}}^*)^{-1})(\widehat{\mathbf{L}}^*\mathbf{x}) = \widehat{\mathbf{L}}^{-1}\mathbf{b}, \tag{8.4}$$

where $\widehat{\mathbf{L}}$ is an appropriate nonsingular matrix, and the unpreconditioned CG (Algorithm 8.1) is formally considered for solving (8.4) instead of $\mathbf{Ax} = \mathbf{b}$. Since the original problem with the matrix $\mathbf{A}$ is in such derivation considered as primary, the subsequent transformation to the algebraic vectors $\mathbf{x}_n$ and $\mathbf{r}_n$ corresponding to this unpreconditioned algebraic system $\mathbf{Ax} = \mathbf{b}$ gives the preconditioned algebraic CG given as Algorithm 6.1, with $\mathbf{M}$ replaced by the algebraic preconditioner

$$\widehat{\mathbf{M}} = \widehat{\mathbf{L}}\widehat{\mathbf{L}}^*; \tag{8.5}$$

see, e.g., [172, 83, 134, 11]. A slightly different description based on the concept of acceleration of some basic underlying iteration scheme can be found in [93, Section 9.4.4]; see also [94]. All these descriptions focus on solving systems of linear algebraic equations separately from the discretization issues. Without considering computational issues such as effects of rounding errors, they are mathematically equivalent. However, as Hackbusch emphasizes in [93, p. 273]:

> *This yields the so-called 'preconditioned CG method' (but notice that the gradients not the CG method are preconditioned),*

which is precisely the point made in the derivation present in Chapter 5.

The purely algebraic view to preconditioning thus includes, without stating it explicitly, a preference for the basis Φ and for the coordinates $\mathbf{x}$ of the solution x_h in this basis, $x_h = \Phi\mathbf{x}$. There might be a reason for such preference coming from physics or geometry of the problem. Then this is typically also taken into account in considerations on preconditioning. Domain decomposition techniques with coarse space components can serve as examples. Using the standard nodal FEM basis, the elements of the algebraic solution $\mathbf{x}$ give the function values of the discretized solution x_h at the discretization nodes. Therefore the nodal basis may be preferred over the other bases. It should be emphasized, however, that the nodal basis argument holds only under the assumption that the algebraic system $\mathbf{Ax} = \mathbf{b}$ is solved exactly, which is not the case in practice.

With (operator preconditioning related) *algebraic preconditioning being interpreted as orthonormalization of the discretization basis*, a possible information which can lead to an appropriate choice of a close-to-orthonormal basis should naturally be taken into account.

There may be no obvious reason for preferring the originally used discretization basis Φ in forming the approximation to x_h using the computed (i.e., *inaccurate*) algebraic coefficients. In other words, there may be no reason for preferring the (approximation of the) algebraic vectors $\mathbf{x}_n$ associated with Φ and $\mathbf{Ax} = \mathbf{b}$ to the (approximation of the) algebraic vectors $\widehat{\mathbf{L}}^*\mathbf{x}_n$ associated with the preconditioned system.[17] We should bear in mind that

---

[17] It is worth noticing that $\mathbf{x}_n$ and $\widehat{\mathbf{L}}^*\mathbf{x}_n$ approximating the algebraic solutions $\mathbf{x}$ and $\widehat{\mathbf{L}}^*\mathbf{x}$, respectively, at the nth step of CG are computed inaccurately due to rounding errors; i.e., they are also only approximated.

we are interested in approximating the *function* x_h, and approximation of the algebraic vector $\mathbf{x}$ solving the discretization system $\mathbf{Ax} = \mathbf{b}$ by some $\mathbf{x}_n$ serves as a tool; it does *not* represent a goal on its own. Since the algebraic coefficients $\mathbf{x}_n$ or $\widehat{\mathbf{L}}^*\mathbf{x}_n$ are discretization-basis dependent, evaluating accuracy of computations in terms of the purely algebraic error $\mathbf{x}_n - \mathbf{x}$ or $\widehat{\mathbf{L}}(\mathbf{x}_n - \mathbf{x})$ can be misleading.

The quantity of interest is the appropriately measured distance of the computed approximation to (the function) x, i.e., the total error. This depends on, in addition to the vectors of algebraic coefficients, the corresponding discretization bases. The spatial distribution of the total error depends on the spatial distribution of both the discretization and algebraic errors, which can be very different; see [147] or [124, Chapter 5]. Comparisons of various preconditioners in literature using a purely algebraic context and test matrices which originate from possibly very different discretizations of various problems show that the link of discretization with preconditioning and evaluation of the error in function spaces is often not taken into account.

In order to complete our argumentation, we need to show how an *arbitrary algebraic preconditioner* $\widehat{\mathbf{M}}$ from (8.4)–(8.5) can be incorporated into the framework of our derivation of Algorithm 6.1; i.e., we have to find a discretization basis $\widehat{\Phi}$ which can be used in the derivation in Chapter 6 instead of Φ to obtain in Algorithm 6.1 the preconditioner $\widehat{\mathbf{M}}$ instead of $\mathbf{M}$.

The ideas about linking PDE discretization with preconditioning are, as mentioned above, certainly not new. Hierarchical basis preconditioning introduced by Yserentant [199, 202, 200, 201] and using the work of Zienkiewicz et al. from 1970 [203] serves as an example of a technique where the described relationship between preconditioning and transformation of the discretization basis has been used in a particular case, with the matrix $(\widehat{\mathbf{L}}^*)^{-1}$ in (8.4) explicitly given as the transformation matrix from the hierarchical to nodal basis; see also [75, Section 11.2], [13], [14]. The algebraic preconditioning is considered there, unlike in Chapters 5 and 6 of our text, as a transformation of the unpreconditioned problem (or method). Analogous ideas (combined with the idea of an appropriate choice of the inner product $(\cdot, \cdot)_{V_h}$ approximating $a(\cdot, \cdot)$; see also Chapter 4) are inherently present in the context of multilevel PDE solvers as well as in the stable subspace splitting and domain decomposition–based techniques; see, e.g., [29, 197, 48, 146, 87], [116, in particular Section 3.4], [158, 152, 81, 2, 186, 100]. The description given herein is general, with preconditioning considered not as a remedy for improving the properties of the obtained (unpreconditioned) algebraic system but rather as an inherent issue which should not be separated from discretization.

In order to demonstrate that the task is subtle and different from, e.g., transformation of the basis in the hierarchical basis preconditioning, we first present a seemingly straightforward approach which, however, does not give the desired result.

Considering the decompositions of the symmetric positive definite matrices

$$\mathbf{M} = \mathbf{LL}^*, \qquad \widehat{\mathbf{M}} = \widehat{\mathbf{L}}\widehat{\mathbf{L}}^*,$$

it is easy to interpret the inner product defined by the matrix $\mathbf{M}$ as the inner product defined by the matrix $\widehat{\mathbf{M}}$ for the transformed vectors. Indeed, substituting $\mathbf{I} = \mathbf{L}^{-1}\mathbf{M}(\mathbf{L}^*)^{-1}$,

$$\widehat{\mathbf{v}}^*\widehat{\mathbf{M}}\widehat{\mathbf{u}} = \widehat{\mathbf{v}}^*\widehat{\mathbf{L}}\widehat{\mathbf{L}}^*\widehat{\mathbf{u}} = \widehat{\mathbf{v}}^*\widehat{\mathbf{L}}(\mathbf{L}^{-1}\mathbf{M}(\mathbf{L}^*)^{-1})\widehat{\mathbf{L}}^*\widehat{\mathbf{u}}$$

$$= ((\mathbf{L}^*)^{-1}\widehat{\mathbf{L}}^*\widehat{\mathbf{v}})^*\mathbf{M}((\mathbf{L}^*)^{-1}\widehat{\mathbf{L}}^*\widehat{\mathbf{u}}) = \mathbf{v}^*\mathbf{Mu},$$

where the transformed vectors corresponding to the coordinates of functions in V in the desired discretization basis $\widehat{\Phi}$ are given by

$$\widehat{\mathbf{v}} = (\widehat{\mathbf{L}}^*)^{-1}\mathbf{L}^*\mathbf{v}, \qquad \widehat{\mathbf{u}} = (\widehat{\mathbf{L}}^*)^{-1}\mathbf{L}^*\mathbf{u}. \tag{8.6}$$

Writing, as above, the matrix $\mathbf{M}$ symbolically as

$$\mathbf{M} = (M_{ij})_{i,j=1}^{N} = ((\phi_j, \phi_i)_{V_h})_{i,j=1}^{N} := \begin{pmatrix} \phi_1^* \\ \vdots \\ \phi_N^* \end{pmatrix} (\phi_1, \ldots, \phi_N) = \Phi^*\Phi$$

gives

$$\widehat{\mathbf{v}}^*\widehat{\mathbf{M}}\widehat{\mathbf{u}} = \mathbf{v}^*\mathbf{M}\mathbf{u} = (\Phi\mathbf{v})^*(\Phi\mathbf{u}) = (\Phi(\mathbf{L}^*)^{-1}\widehat{\mathbf{L}}^*\widehat{\mathbf{v}})^*(\Phi(\mathbf{L}^*)^{-1}\widehat{\mathbf{L}}^*\widehat{\mathbf{u}})$$
$$= \widehat{\mathbf{v}}^*\widehat{\Phi}^*\widehat{\Phi}\widehat{\mathbf{u}}.$$

This leads to

$$\widehat{\mathbf{M}} = \widehat{\Phi}^*\widehat{\Phi}, \qquad \text{i.e.,} \qquad \widehat{\mathbf{M}} = (\widehat{M}_{ij})_{i,j=1,\ldots,N} = ((\widehat{\phi}_j, \widehat{\phi}_i)_{V_h})_{i,j=1,\ldots,N}, \tag{8.7}$$

where

$$\widehat{\Phi} = \Phi(\mathbf{L}^*)^{-1}\widehat{\mathbf{L}}^* := (\widehat{\phi}_1, \ldots, \widehat{\phi}_n) \quad \text{with} \quad \widehat{\phi}_\ell = \Phi((\mathbf{L}^*)^{-1}\widehat{\mathbf{L}}^*)\mathbf{e}_\ell, \quad \ell = 1, \ldots, N. \tag{8.8}$$

Consequently, we observe that taking formally any (symmetric positive definite) preconditioning $\widehat{\mathbf{M}}$ replacing $\mathbf{M}$ in Algorithm 6.1 corresponds within the given CG derivation to transformation of the original discretization basis Φ to the transformed basis $\widehat{\Phi}$ given by (8.8). The matrix representation of the Riesz map $\widehat{\tau}$, which can be associated with the matrix $\widehat{\mathbf{M}}$ and the transformed basis $\widehat{\Phi}$, is then given by

$$\widehat{\tau}\widehat{\Phi}^{\#} = \widehat{\Phi}\widehat{\mathbf{M}}^{-1}, \tag{8.9}$$

where the transformation matrix of the canonical dual basis is given, as above, by standard linear algebra as the inverse transpose of the transformation matrix for the primal basis,

$$\widehat{\Phi}^{\#} = \Phi^{\#}\mathbf{L}\widehat{\mathbf{L}}^{-1} := (\widehat{\phi}_1^{\#}, \ldots, \widehat{\phi}_n^{\#}) \quad \text{with} \quad \widehat{\phi}_\ell^{\#} = \Phi^{\#}(\mathbf{L}\widehat{\mathbf{L}}^{-1})\mathbf{e}_\ell, \quad \ell = 1, \ldots, N. \tag{8.10}$$

Indeed, with (8.10) and denoting $\mathbf{c}_j = (\mathbf{L}^*)^{-1}\widehat{\mathbf{L}}^*\mathbf{e}_j := (\xi_1^j, \ldots, \xi_N^j)^*$ the jth column of the transformation matrix $(\mathbf{L}^*)^{-1}\widehat{\mathbf{L}}^*$ and $\mathbf{d}_i = \mathbf{L}\widehat{\mathbf{L}}^{-1}\mathbf{e}_i := (\eta_1^i, \ldots, \eta_N^i)^*$ the ith column of the transformation matrix $\mathbf{L}\widehat{\mathbf{L}}^{-1}$, we get

$$\langle \widehat{\phi}_i^{\#}, \widehat{\phi}_j \rangle = \langle \Phi^{\#}\mathbf{d}_i, \Phi\mathbf{c}_j \rangle = \sum_{\ell=1}^{N} \eta_\ell^i \xi_\ell^j = \mathbf{c}_j^*\mathbf{d}_i = \mathbf{e}_j\widehat{\mathbf{L}}\mathbf{L}^{-1}\mathbf{L}\widehat{\mathbf{L}}^{-1}\mathbf{e}_i = \delta_{ij};$$

i.e., $\widehat{\Phi}^{\#}$ is the canonical dual basis to $\widehat{\Phi}$. Substituting for $\widehat{\Phi}^{\#}$, $\widehat{\Phi}$, and $\widehat{\mathbf{M}}$ in (8.9) gives after easy manipulations

$$\widehat{\tau}\Phi^{\#} = \Phi\mathbf{M}^{-1}; \tag{8.11}$$

i.e., the matrix representation of the transformed Riesz map $\widehat{\tau}$ using the original basis functions $\Phi^{\#}$ and Φ is identical to the matrix representation of the original Riesz map using the same basis functions. This makes sense, since the inner product $(\cdot, \cdot)_{V_h}$ remains unchanged and transformation of the discretization basis does not change the Riesz operator.

Of course, transformation of the discretization basis means also transformation of the coordinate vectors and of the matrices in Algorithm 6.1. Therefore with

$$\widehat{\Phi} = \Phi(\mathbf{L}^*)^{-1}\widehat{\mathbf{L}}^*, \qquad \widehat{\Phi}^{\#} = \Phi^{\#}\mathbf{L}\widehat{\mathbf{L}}^{-1}$$

leading to replacing $\mathbf{M}$ with $\widehat{\mathbf{M}}$ in Algorithm 6.1, we have also to replace $\mathbf{A}$ with $\widehat{\mathbf{A}}$ obtained from

$$\mathcal{A}\,w = \mathcal{A}\,\Phi\mathbf{w} = \mathcal{A}(\widehat{\Phi}(\widehat{\mathbf{L}}^*)^{-1}\mathbf{L}^*)\mathbf{w} = \mathcal{A}\widehat{\Phi}((\widehat{\mathbf{L}}^*)^{-1}\mathbf{L}^*\mathbf{w}) = \mathcal{A}\,\widehat{\Phi}\widehat{\mathbf{w}}$$
$$= \Phi^{\#}\mathbf{A}\mathbf{w} = (\widehat{\Phi}^{\#}\widehat{\mathbf{L}}\mathbf{L}^{-1})\mathbf{A}\mathbf{w}$$
$$= \widehat{\Phi}^{\#}(\widehat{\mathbf{L}}\mathbf{L}^{-1}\mathbf{A}(\mathbf{L}^*)^{-1}\widehat{\mathbf{L}}^*)((\widehat{\mathbf{L}}^*)^{-1}\mathbf{L}^*\mathbf{w})$$
$$= \widehat{\Phi}^{\#}\widehat{\mathbf{A}}\widehat{\mathbf{w}},$$

i.e.,

$$\widehat{\mathbf{A}} = \widehat{\mathbf{L}}\mathbf{L}^{-1}\mathbf{A}(\mathbf{L}^*)^{-1}\widehat{\mathbf{L}}^*.$$

Considering further, for any $f \in V_h^{\#}$,

$$f = \Phi^{\#}\mathbf{f} = \widehat{\Phi}^{\#}\widehat{\mathbf{f}},$$

we have to replace $\mathbf{b}$ with $\widehat{\mathbf{b}} = \widehat{\mathbf{L}}\mathbf{L}^{-1}\mathbf{b}$, $\mathbf{r}_n$ with $\widehat{\mathbf{r}}_n = \widehat{\mathbf{L}}\mathbf{L}^{-1}\mathbf{r}_n$. Using the transformation of the vectors corresponding to the functions in V_h described above (see (8.6)) in derivation of $\widehat{\Phi}$, we finally replace $\mathbf{x}_n$ with $\widehat{\mathbf{x}}_n = (\widehat{\mathbf{L}}^*)^{-1}\mathbf{L}^*\mathbf{x}_n$, $\widehat{\mathbf{p}}_n = (\widehat{\mathbf{L}}^*)^{-1}\mathbf{L}^*\mathbf{p}_n$, and $\widehat{\mathbf{z}}_n = (\widehat{\mathbf{L}}^*)^{-1}\mathbf{L}^*\mathbf{z}_n$, which completes the argument. (With $\widehat{\mathbf{M}} = \mathbf{I}$ we get (8.3) with the interpretation as above.)

The obtained result is, however, unsatisfactory, because $\widehat{\mathbf{A}}$ is for $\widehat{\mathbf{L}} \neq \mathbf{L}$ different from $\mathbf{A}$, and similarly, $\widehat{\mathbf{b}}, \widehat{\mathbf{x}}, \widehat{\mathbf{x}}_n, \widehat{\mathbf{p}}_n, \widehat{\mathbf{r}}_n$, and $\widehat{\mathbf{z}}_n$ are different from $\mathbf{b}, \mathbf{x}, \mathbf{x}_n, \mathbf{p}_n, \mathbf{r}_n$, and $\mathbf{z}_n$, respectively. We will therefore return to (8.4) and explain that in order to interpret algebraic preconditioning as transformation of the discretization basis, we must simultaneously also transform the inner product $(\cdot,\cdot)_{V_h}$ used in discretization (i.e., we must change the Riesz map τ).

In order to get an algebraic preconditioning using $\widehat{\mathbf{M}} \neq \mathbf{M}$, where $\mathbf{M}$ is determined by Φ and the inner product $(\cdot,\cdot)_{V_h}$ via $\mathbf{M} = \Phi^*\Phi$, we consider a discretization of the infinite-dimensional CG with the discretization bases

$$\widehat{\Phi} = \Phi(\widehat{\mathbf{L}}^*)^{-1}, \quad \widehat{\Phi}^{\#} = \Phi^{\#}\widehat{\mathbf{L}}. \tag{8.12}$$

With this choice

$$\mathcal{A}\widehat{\Phi} = \widehat{\Phi}\widehat{\mathbf{A}}$$

would give (analogously to $\mathbf{A}_t$ above) $\widehat{\mathbf{A}} = \widehat{\mathbf{L}}^{-1}\mathbf{A}(\widehat{\mathbf{L}}^*)^{-1}$. Multiplication of $\mathbf{x}$ by $\widehat{\mathbf{L}}^*$ and of the whole equation by $\widehat{\mathbf{L}}^{-1}$ represents the associated transformation of the functions in V_h as

$$\widehat{\mathbf{u}} = \widehat{\mathbf{L}}^*\mathbf{u}$$

and the functionals in $V_h^{\#}$ as

$$\widehat{\mathbf{f}} = \widehat{\mathbf{L}}^{-1}\mathbf{f},$$

respectively. In this way, we would inevitably get an additional nontrivial preconditioning resulting from the discretization using $\widehat{\Phi}$. Indeed, for $\widehat{L} \neq L$ (with no loss of generality we do not consider here a trivial scaling of the columns of $\widehat{L}$ by ± 1) the basis $\widehat{\Phi}$ cannot be orthonormal with respect to $(\cdot, \cdot)_{V_h}$:

$$\widehat{\Phi}^* \widehat{\Phi} = \widehat{L}^{-1} \Phi^* \Phi (\widehat{L}^*)^{-1} = \widehat{L}^{-1} L L^* (\widehat{L}^*)^{-1}.$$

In order to achieve the orthonormality of $\widehat{\Phi}$, we must change the inner product in V_h.

The inner product $(\cdot, \cdot)_{V_h}$ in Chapter 6 (see (6.2)) is given by

$$(u, v)_{V_h} = (\Phi u, \Phi v)_{V_h} = v^* M u,$$

where u, v are coordinates of u and v in the basis Φ, respectively, i.e.,

$$u = \Phi u, \qquad u = (\Phi^\# u)^*,$$
$$v = \Phi v, \qquad v = (\Phi^\# v)^*.$$

In order to obtain the interpretation of the algebraic preconditioning $\widehat{M}$ as transformation of the basis given by (8.12),

$$\Phi \to \widehat{\Phi}, \quad \widehat{\Phi} = \Phi (\widehat{L}^*)^{-1},$$

i.e., in order to obtain with the discretization basis $\widehat{\Phi} = \Phi (\widehat{L}^*)^{-1}$ the unpreconditioned CG applied to the preconditioned algebraic system (8.4), we must consider together with the transformation basis $\widehat{\Phi}$ the inner product in V_h defined by

$$(u, v)_{\text{new}, V_h} = (\widehat{\Phi} \widehat{u}, \widehat{\Phi} \widehat{v})_{\text{new}, V_h} := \widehat{v}^* \widehat{u} = v^* \widehat{L} \widehat{L}^* u = v^* \widehat{M} u. \tag{8.13}$$

This means that the arbitrarily given algebraic preconditioning is interpreted as a transformation of the discretization basis and, at the same time, a transformation of the inner product in V_h such that the transformed basis $\widehat{\Phi}$ is orthonormal with respect to the transformed inner product. Indeed,

$$(\widehat{\phi}_i, \widehat{\phi}_j)_{\text{new}, V_h} = (\widehat{\Phi} e_i, \widehat{\Phi} e_j)_{\text{new}, V_h} = e_j^* e_i = \delta_{ij}$$

and the matrix representation of the Riesz map $\widehat{\tau}$ defined by the transformed inner product $(\cdot, \cdot)_{\text{new}, V_h}$ is given by

$$\widehat{\tau} \widehat{\Phi}^\# = \widehat{\Phi} M_{\widehat{\tau}},$$

where from

$$\widehat{v}^* \widehat{f} = \langle f, \widehat{v} \rangle = (\widehat{\tau} f, \widehat{v})_{\text{new}, V_h} = (\widehat{\tau} \widehat{\Phi}^\# f, \widehat{\Phi} \widehat{v})_{\text{new}, V_h} = (\widehat{\Phi} M_{\widehat{\tau}} \widehat{f}, \widehat{\Phi} \widehat{v})_{\text{new}, V_h}$$
$$= \widehat{v}^* M_{\widehat{\tau}} \widehat{f},$$

we get, analogously to the derivation of (6.5), $M_{\widehat{\tau}} = I$. This completes the argument.

Chapter 9

Fundamental Theorem on Discretization

Discretization in Chapter 7 as well as preconditioning of the linear algebraic problem described as transformation of the discretization basis (and the associated transformation of the inner product) in Chapter 8 were presented without any notion of geometric substructuring of the solution domain Ω using meshes or decompositions of the domain. Practical use of numerical methods in PDEs requires, however, construction of finite-dimensional subspaces V_h in a systematic way, which leads, as the main example, to the highly developed mathematical technology of piecewise polynomial approximations and the FEM; see, e.g., [30, Chapter 3], [45], or [76, Chapter 4]. This chapter recalls consistency, stability and convergence of discretization schemes as presented in [6] and [9].

For the moment we will slightly generalize our settings from the last chapters and consider bounded and coercive linear operators $\mathscr{A} : V \to V^{\#}$, with the associated bounded bilinear forms $a(\cdot,\cdot) : V \times V \to \mathbb{R}$. Then the equation (1.1) (or equivalently, (3.4) or (3.5)) has for any $b \in V^{\#}$ the unique solution $u \in V$. The fundamental theorem on discretization schemes parametrized by a positive discretization parameter h states that a *discretization scheme which is consistent and stable is convergent*; see [6, 9]. Here *consistency* describes in an appropriate (problem dependent) way how closely the discrete problem $\mathscr{A}_h u_h = b_h$ defined by $\mathscr{A}_h$ and b_h approximates the original problem $\mathscr{A} u = b$ defined by $\mathscr{A}$ and b. It should be understood, as stated in [9, p. 282], that

> *consistency is not and cannot be defined to mean norm convergence of the discrete operators to the PDE operator, since the PDE operator, being an invertible operator between infinite-dimensional spaces, is not compact and so is not the norm limit of the finite-dimensional operators.*[18]

The appropriate way of quantifying consistency of the discretization scheme is to ask to which extent the discrete equation is satisfied by the true solution. The consistency error can simply be measured by the *residual*

$$\mathscr{A}_h \pi_h u - b_h, \tag{9.1}$$

where $\pi_h u$ denotes an appropriate representative of the solution u in V_h. This representative can be constructed in various ways, appropriately chosen for the given discretization scheme; e.g., by some restriction or interpolation. (We will comment on the particular

---

[18]For the corresponding statements on compact operators we refer to, e.g., [111], [46, p. 174], [54, p. 486] or [10, p. 98].

construction of $\pi_h u$ relevant to the Galerkin discretization later.) This immediately gives

$$\pi_h u - u_h = \mathscr{A}_h^{-1}(\mathscr{A}_h \pi_h u - b_h), \tag{9.2}$$

with the bound (using some norm $\|\cdot\|_h$ in V_h and $\|\cdot\|_{h^\#}$ in $V_h^\#$)

$$\|\pi_h u - u_h\|_h \leq \left(\sup_{z \in V_h^\#; \|z\|_{h^\#} = 1} \|\mathscr{A}_h^{-1} z\|_h \right) \|\mathscr{A}_h \pi_h u - b_h\|_{h^\#}. \tag{9.3}$$

If the discretization scheme is *stable*[19], i.e., if the mappings $\mathscr{A}_h^{-1} : V_h^\# \to V_h$ are continuous with the norm bounded uniformly in h, then the *consistency* of the discretization scheme $\|\mathscr{A}_h \pi_h u - b_h\|_{h^\#} \to 0$ with $h \to 0$ gives the *convergence*

$$\|\pi_h u - u_h\|_h \to 0 \quad \text{as} \quad h \to 0.$$

We will now consider the Galerkin method where the discretization space V_h is a subspace of V. Then the best approximation $\pi_h u \in V_h$ to the solution $u \in V$ of the functional equation $\mathscr{A} u = b$ can be given by the orthogonal projection π_h onto V_h using the appropriately chosen inner product $(\cdot,\cdot)_G$. This inner product defines the orthogonality which determines the meaning of the *optimal* (best) approximation (this inner product is not necessarily equal or related to any of the inner products $(\cdot,\cdot)_V$ or $(\cdot,\cdot)_a$ which are used in Chapter 5 in derivation of CG). Then, using the associated norm $\|\cdot\|_{V_G}$, we have the best approximation property

$$\|u - \pi_h u\|_{V_G} = \inf_{v \in V_h} \|u - v\|_{V_G}, \tag{9.4}$$

which is immediately visible from the following equality:

$$\|u - v\|_{V_G}^2 = \|u - \pi_h u + \pi_h u - v\|_{V_G}^2 = \|u - \pi_h u\|_{V_G}^2 + \|\pi_h u - v\|_{V_G}^2,$$

which holds for all $v \in V_h$. The property (9.4) is given by the inclusion $V_h \subset V$ and by the choice of $(\cdot,\cdot)_G$. It need not be related to the Galerkin orthogonality $\langle \mathscr{A}_h u_h - b_h, v \rangle = 0$ for all $v \in V_h$, which determines the discretization.

Within this Galerkin setting and using the notation analogous to that in Chapters 3 and 4 with $\mathscr{A}$ bounded and α_G-coercive with respect to the norm $\|\cdot\|_{V_G}$,

$$\sup_{z \in V_h^\#; \|z\|_{V_G^\#} = 1} \|\mathscr{A}_h^{-1} z\|_{V_G} \leq \sup_{z \in V^\#; \|z\|_{V_G^\#} = 1} \|\mathscr{A}^{-1} z\|_{V_G} = 1/\alpha_G,$$

i.e., the stability of the Galerkin method with respect to the norm $\|\cdot\|_{V_G}$ is automatic, and also

$$\sup_{w \in V_h; \|w\|_{V_G} = 1} \|\mathscr{A}_h w\|_{V_G^\#} \leq \sup_{w \in V; \|w\|_{V_G} = 1} \|\mathscr{A} w\|_{V_G^\#} = C_G.$$

Therefore (9.3) gives

$$\|\pi_h u - u_h\|_{V_G} \leq \alpha_G^{-1} \|\mathscr{A}_h \pi_h u - b_h\|_{V_G^\#}. \tag{9.5}$$

---

[19]Alternatively, the discretization scheme is *stable* if the discrete problem possesses a unique solution with

$$\|u_h\|_h = \|\mathscr{A}_h^{-1} b_h\|_h \leq \frac{\|b_h\|_{h^\#}}{\alpha_h} \qquad \text{for all } b_h \in V_h^\#$$

and there is an $\alpha_0 > 0$ such that $1/\alpha_h \leq 1/\alpha_0$ for all $h > 0$.

Using (7.3) we additionally have

$$
\begin{aligned}
\|\mathscr{A}_h \pi_h u - b_h\|_{V_G^{\#}} &= \sup_{v \in V_h; \|v\|_{V_G}=1} |\langle \mathscr{A}_h \pi_h u - b_h, v \rangle| \\
&= \sup_{v \in V_h; \|v\|_{V_G}=1} |\langle \mathscr{A} \pi_h u - b, v \rangle| = \sup_{v \in V_h; \|v\|_{V_G}=1} |\langle \mathscr{A}(\pi_h u - u), v \rangle| \\
&\leq \sup_{z \in V; \|z\|_{V_G}=1} \|\mathscr{A} z\|_{V_G^{\#}} \|\pi_h u - u\|_{V_G}.
\end{aligned}
\tag{9.6}
$$

In combination with (9.5), (9.4), and an analogue of (4.7) for $\|\cdot\|_{V_G}$, C_G, and α_G,

$$
\|\pi_h u - u_h\|_{V_G} \leq \varkappa_G(\mathscr{A}) \inf_{v \in V_h} \|u - v\|_{V_G}.
$$

With the triangle inequality $\|u - u_h\|_{V_G} \leq \|u - \pi_h u\|_{V_G} + \|u_h - \pi_h u\|_{V_G}$ this finally results in the bound

$$
\|u - u_h\|_{V_G} \leq (1 + \varkappa_G(\mathscr{A})) \inf_{v \in V_h} \|u - v\|_{V_G}.
\tag{9.7}
$$

Here we have intentionally used the general derivation linking convergence of the discretization scheme to its consistency and stability. As pointed out in [9, p. 243], different arguments in [198] allow avoiding the presence of 1 in (9.7). More precisely, the authors of [198] construct the linear operator $P_h : V \to V_h$ such that $\|P_h\|_{\mathscr{L}(V;V)} = \|I - P_h\|_{\mathscr{L}(V;V)}$ and for any $z \in V$, where $\mathscr{A} z = g$ for g defined as $g := \mathscr{A} z$, $P_h z = z_h$, where z_h solves the discretized problem $\mathscr{A}_h z_h = g_h$. Then for $v \in V_h$ (i.e., $P_h v = v$)

$$
\begin{aligned}
\|u - u_h\|_{V_G} &= \|(I - P_h)(u - v)\|_{V_G} \leq \|I - P_h\|_{\mathscr{L}(V;V)} \|u - v\|_{V_G} \\
&= \|P_h\|_{\mathscr{L}(V;V)} \|u - v\|_{V_G}.
\end{aligned}
$$

Since for any $z \in V$

$$
\begin{aligned}
\|z_h\|_{V_G}^2 &\leq \alpha_G^{-1} \langle \mathscr{A} z_h, z_h \rangle \leq \alpha_G^{-1} \|z_h\|_{V_G} \sup_{v \in V_h; \|v\|_{V_G}=1} \langle \mathscr{A} z_h, v \rangle \\
&= \alpha_G^{-1} \|z_h\|_{V_G} \sup_{v \in V_h; \|v\|_{V_G}=1} \langle \mathscr{A} z, v \rangle \leq \alpha_G^{-1} C_G \|z_h\|_{V_G} \|z\|_{V_G},
\end{aligned}
$$

we get $\|P_h z\|_{V_G} = \|z_h\|_{V_G} \leq \varkappa_G(\mathscr{A}) \|z\|_{V_G}$, which implies $\|P_h\|_{\mathscr{L}(V;V)} \leq \varkappa_G(\mathscr{A})$ and the assertion (9.7) without 1 follows.

If $\mathscr{A}$ is bounded and α_G-coercive, as assumed throughout the text, then a simpler derivation (based on the Galerkin orthogonality (7.6)) yields

$$
\begin{aligned}
\alpha_G \|u - u_h\|_{V_G}^2 &\leq a(u - u_h, u - u_h) \\
&= a(u - u_h, u - v) + a(u - u_h, v - u_h) \qquad (v \in V_h) \\
&= a(u - u_h, u - v) \\
&\leq C_G \|u - u_h\|_{V_G} \|u - v\|_{V_G},
\end{aligned}
$$

which gives the same result,

$$
\|u - u_h\|_{V_G} \leq \varkappa_G(\mathscr{A}) \inf_{v \in V_h} \|u - v\|_{V_G},
\tag{9.8}
$$

known as the Cea lemma; see [30, Section 2.8] or [159, Section 12.4.4]. If $\mathscr{A}$ is also self-adjoint (i.e., the associated bilinear form $a(\cdot, \cdot)$ is symmetric), then it is natural to define optimality of the approximation in V_h in terms of the energy norm $\|\cdot\|_a$; see Chapter 5. With this choice $(\cdot,\cdot)_G = (\cdot,\cdot)_a$, $u_h = \pi_h u$, the Cea lemma is trivial and

$$\|u - u_h\|_a = \inf_{v \in V_h} \|u - v\|_a. \tag{9.9}$$

With different choices $(\cdot,\cdot)_G \neq (\cdot,\cdot)_a$ the Cea lemma links in fact through the constants α_G and C_G the optimal $\|u - u_h\|_a$ with some other approximation error, for example $\|u - u_h\|_V$ or $\|u - u_h\|_{V_G}$, and one gets (using (9.9)) that

$$\alpha_G \|u - u_h\|_{V_G}^2 \leq a(u - u_h, u - u_h) = \|u - u_h\|_a^2 = \left(\inf_{v \in V_h} \|u - v\|_a \right)^2$$

$$\leq C_G \left(\inf_{v \in V_h} \|u - v\|_{V_G} \right)^2,$$

leading to

$$\|u - u_h\|_{V_G} \leq \sqrt{\varkappa_G(\mathscr{A})} \inf_{v \in V_h} \|u - v\|_{V_G}, \tag{9.10}$$

which improves (9.8).

The Cea lemma is important when $\mathscr{A}$ is not self-adjoint and $a(\cdot,\cdot)$ does not represent an inner product. In such cases optimality of the approximation in V_h cannot be easily identified with the physics of the problem and this represents the true difficulty in numerical analysis of such problems. We can observe one of the points where our restriction to self-adjoint, bounded, and α-coercive linear operators (and to the CG method on the computational side) makes the analysis easy. Possible generalizations will have to use a different approach here.

We see that although (9.8) gives the bound for the discretization error proportional to the best approximation to the solution in the discretization subspace (both measured in the same norm), the constant of proportionality is not benign. If $\varkappa_G(\mathscr{A})$ is large, then the Galerkin approximation u_h is not proved to be close to the best approximation to u from V_h.

One may notice the following observation. While u and u_h depend on the right-hand side b of the functional equation, and the concept of consistency measures the quality of the operator and data discretization via the residual $\mathscr{A}_h \pi_h u - b_h$, the fact that we use in (9.3), (9.5), and also in (9.6) the upper bounds

$$\sup_{w \in V; \|w\|_{V_G} = 1} \|\mathscr{A} w\|_{V_G^{\#}} \leq C_G \quad \text{and} \quad \sup_{z \in V^{\#}; \|z\|_{V_G^{\#}} = 1} \|\mathscr{A}^{-1} z\|_{V_G} \leq 1/\alpha_G$$

for the norms of $\mathscr{A}_h$ and $\mathscr{A}_h^{-1}$ leads to the multiplicative factor $\varkappa_G(\mathscr{A})$ in (9.8), which is *independent of b* and therefore represents *the worst case multiplicative factor with respect to b*. This is not harmful *here*, but an analogous approach is harmful whenever used within the quantitative evaluation of the computational cost requiring a posteriori error estimates.

One may notice an analogy with solving the linear algebraic system $\mathbf{Ax} = \mathbf{b}$, where the simple bound for the size of the error $\mathbf{x} - \mathbf{y}$, where $\mathbf{y}$ is an approximation to $\mathbf{x}$, is given in terms of the residual $\mathbf{r} = \mathbf{b} - \mathbf{Ay}$ by (see, e.g., [97, Chapter 7])

$$\frac{\|\mathbf{x} - \mathbf{y}\|}{\|\mathbf{x}\|} = \frac{\|\mathbf{A}^{-1}\mathbf{r}\|}{\|\mathbf{x}\|} \leq \frac{\|\mathbf{A}^{-1}\| \, \|\mathbf{A}\| \, \|\mathbf{r}\|}{\|\mathbf{A}\| \, \|\mathbf{x}\|} \leq \varkappa(\mathbf{A}) \frac{\|\mathbf{r}\|}{\|\mathbf{b}\|}.$$

Here $\|\cdot\|$ is any vector and the subordinate matrix norm. For an ill-conditioned matrix $\mathbf{A}$, a small residual $\mathbf{r}$ does not guarantee a small error $\mathbf{x} - \mathbf{y}$. On the other hand, in fortunate cases it *may* happen that the error $\mathbf{x} - \mathbf{y}$ is small even when $\varkappa(\mathbf{A})$ is large.

The importance of keeping conditioning of the algebraic problems under control for numerical reasons was noticed a long time ago; see, e.g., the description of the link between possible numerical error due to roundoff and the ill-conditioning in [47, Section 1.8 and Chapter 18]. That text also discusses possible sources of ill-conditioning and how to avoid them. Kannan, Hendry, Higham, and Tisseur recently published a very insightful paper [102] on this topic in which they show how mistakes in modeling and discretization which lead to ill-conditioned problems can be detected numerically. More precisely, they present a method to identify elements and degrees of freedom in the given FEM discretization that cause ill-conditioning in the algebraic problem. Their method is based on the investigation and numerical calculation of the eigenvectors of the matrix $\mathbf{A}$ corresponding to its extremal eigenvalues, and it represents a beautiful example of numerical analysis.

Chapter 10

Local and Global Information in Discretization and in Computation

Any discretization scheme has to deal with the discretization error

$$u - u_h \quad \text{where} \quad \mathscr{A}_h u_h = b_h.$$

The question on how to measure the *size of the error* $u - u_h$ is of fundamental importance. In a *qualitative* reasoning on convergence, including a priori FEM error bounds used for that purpose, one can utilize the (topological) equivalence of norms; see also Section 2.3. In evaluation of *accuracy and cost of numerical computations*, however, an appropriate measure of the size of $u - u_h$ (choice of an appropriate norm) must reflect the goal of the computation and is determined by the problem to be solved. This issue is nicely formulated, e.g., by Babuška and Strouboulis in [16, p. 417]:

> *In engineering practice it is not sufficient to estimate only the energy norm of the error because a small value of the global energy norm of the error does not necessarily imply that the error in the outputs of interest is also small (e.g., a 5% relative error in the global energy norm does not imply 5% relative error in the maximum stress in a region of interest). [...] An essential requirement is that the quantity of interest has to be well defined; for example, it is meaningless to ask for an estimate of the maximum error in the derivative, flux, or stress for a problem set in a polygonal non-convex domain, because the exact value does not exist (the derivative, flux, or stress in the neighborhood of a corner point is usually unbounded).*

Another clear point is made in the introduction of the paper [73] by Giles and Süli:

> *In many scientific and engineering applications [...] the objective is merely a rough, qualitative assessment of the details of the analytical solution over the computational domain, the quantitative concern being directed towards a few output functionals, derived quantities of particular engineering or scientific relevance.*

In [18] it is moreover emphasized that the Galerkin solution u_h is not known, and the error associated with numerical *computations* must not be excluded from considerations; see also the quote from that paper presented in the next chapter. This particular issue is addressed in some more detail in Chapter 12 below, with additional information presented in [3, Chapter 4], [124, Chapter 5], [101, 147], and in the references given there.

In standard Galerkin FEM discretizations the key paradigm is construction of bases for the discrete spaces V_h (parametrized by the discretization parameter h) consisting of functions with *local supports*. More specifically, the set Ω is decomposed into simplexes, and for each simplex there is a space of polynomials called the shape functions, and a set of degrees of freedom for the shape functions, with each degree of freedom associated to a face of some dimension of the simplex. This allows for efficient implementation of the operations in the *global* piecewise polynomial subspace; see [9, Section 1.1]. The efficient implementation of the FEM approximations technology is built on *locality of the basis functions*, i.e., on the fact that the functions generating the finite-dimensional spaces V_h vanish on all but a small number of simplexes determining the decomposition of Ω. The resulting sparsity of the matrix representation $\mathbf{A}$ of the discretized operator $\mathscr{A}_h$ is considered a major *computational advantage* of the FEM discretizations; see, e.g., [75, Section 4.1.2] and also the discussion in [76, Section 11.4]. We will look at this issue in more detail.

First, consider the FEM representation of the discretized operator $\mathscr{A}_h$ by a sparse matrix $\mathbf{A} = (A_{ij})_{i,j=1,\dots,N} = (a(\phi_j, \phi_i))_{i,j=1,\dots,N}$; see Chapter 8. Solving the linear algebraic equation $\mathbf{A}\mathbf{x} = \mathbf{b}$ iteratively means construction of a sequence $\mathbf{x}_0, \mathbf{x}_1, \dots$ (starting from an initial guess $\mathbf{x}_0$, often $\mathbf{x}_0 = 0$) such that its terms approximate the true solution $\mathbf{x} = \mathbf{A}^{-1}\mathbf{b}$. Apart from trivial cases, $\mathbf{A}^{-1}$ is *not sparse*, and the individual components of $\mathbf{x}$ substantially depend, in general, on all or on most of the components of $\mathbf{b}$. Assume that the approximation $\mathbf{x}_n$ is constructed at the step n $(n = 1, 2, \dots)$ using the previous approximations $\mathbf{x}_0, \dots, \mathbf{x}_{n-1}$ and no more than one matrix-vector multiplication, which is the case, e.g., in CG and other Krylov subspace methods. (For detailed discussion see [124, Section 4.1].) Then it may take many iterations to approximate the global transfer of information from the individual elements of $\mathbf{b}$ to the individual elements of $\mathbf{A}^{-1}\mathbf{b} = \mathbf{x}$ by the *local information transfers* represented by the multiplications $\mathbf{Ab}$, $\mathbf{A}(\mathbf{Ab})$, ... In other words, let substantial information affecting the solution $\mathbf{x}$ be given by the data associated with points on one side of the domain. Due to the locality of the discretization basis functions $\phi_1, \dots, \phi_N$, this information is associated with only a small number of them. Consequently, using the sparse matrix representation of the discretization problem, it takes many matrix-vector multiplications until this information reaches a distant part of the domain. This is a direct consequence of the locality of the basis functions and of the induced sparsity of the matrix $\mathbf{A}$. Preconditioning is then needed in order to cure the computational inefficiency to which the local FEM discretization has contributed. Viewing preconditioning as transformation of the discretization basis, which enables one to take care of the global information transfer in a fast way, therefore naturally applies to *any* preconditioning and not only to techniques where the coarse space components are explicitly constructed; see, e.g., in addition to references from Chapter 8, [81, 65, 64, 113, 106, 107, 52, 39].

Second, if the linear algebraic problem $\mathbf{A}\mathbf{x} = \mathbf{b}$ resulting from the discretizations is solved *exactly*, then the issue whether the discretization basis is composed only of locally supported basis functions or also of globally supported basis functions affects only the discretization error. But *we do not compute exactly*, and this indisputable fact is of increasing importance in solving challenging problems and, in particular, considering future computer architectures and measuring the computational efficiency not by the operations count or by the execution time but by the total power consumption. The locality of the FEM basis functions then highlights the importance of the following point:

There is no guarantee whatsoever that algebraic errors in numerical approximation of the algebraic solution $\mathbf{x}$ are distributed evenly throughout the individual components of $\mathbf{x}$. The sparsity of the FEM bases means that the spatial distribution of the algebraic error can exhibit large local components, and its spatial distribution over the domain can be very different from the spatial distribution of the discretization error.

When the problem to be solved reflects by its nature a global information shared over the domain, discretizations using only locally supported functions may cause difficulties at the computational stage of the whole solution process. Preconditioning can then be viewed, at least to some extent, as a tool needed to resolve the difficulty to which the computationally inappropriate discretization has contributed. Including the coarse space components in domain decomposition-based techniques offers excellent examples of dealing with this issue at the discretization stage.

Chapter 11

Limits of the Condition Number–Based Descriptions

In several places throughout the text we have encountered "conditioning," either in terms of the quantitative characteristics $\varkappa(\mathscr{A})$ of an operator or $\varkappa(\mathbf{A})$ of a matrix or in a suggestive term "preconditioning." By preconditioning one understands a transformation of the problem to be solved with the purpose of achieving faster convergence behavior of an iterative method and reducing the overall computational cost. While in classical linear stationary iterative methods their asymptotic behavior can be meaningfully described by a single numerical characteristic such as the spectral radius of the iteration matrix or the condition number of the system matrix $\mathbf{A}$, etc., and the same is also true for the Chebyshev semi-iterative method, with Krylov subspace methods the situations is radically different. As clearly formulated by Becker, Johnson and Rannacher in [18]:

> *For large size problems, in particular in three dimensions, direct methods are too work-intensive and only iterative methods such as multigrid methods or preconditioned conjugate gradient type methods may be used. ... Usually, ad hoc stopping criteria are used, e.g., requiring an initial (algebraic) residual to be reduced by a certain ad hoc factor, but these criteria have no clear connection to the actual error in the corresponding approximate solution, which is the quantity of interest. This leaves the user of iterative solutions methods in a serious dilemma: With no objective stopping criterion available, one has either to continue the iterations until the discrete solution error is practically "zero," which increases the computational cost with possibly no gain in the overall precision, or take the risk of stopping the iterations prematurely. In the first case there would be a loss of efficiency and in the second a loss of reliability. ... A solution to this problem can only be obtained by combining aspects of the underlying partial differential equations and the corresponding finite element discretization with aspects of the iterative discrete solution algorithm. A "pure" numerical linear algebra point of view, for instance based on the condition number of the stiffness matrix, does not appear to be able to lead to a balance of discretization and solution errors.*

We have observed in Section 5.2 that CG can be viewed as matrix formulation of the Gauss–Christoffel quadrature determined by the spectral decomposition of the operator $\mathscr{A}$ and the associated decomposition of the initial residual r_0. More generally, Krylov subspace methods can be associated with the Vorobyev moment problem [192]. As explained above, they are *not* linear at all. As an example, computing (assuming exact arithmetic) the nth CG approximation x_n to the solution x in fact means solving the following $2n$

nonlinear equations for the unknowns $\theta_1^{(n)}, \ldots, \theta_n^{(n)}, \omega_1^{(n)}, \ldots, \omega_n^{(n)}$ (see (5.31)):

$$\sum_{j=1}^{n} \{\theta_j^{(n)}\}^\ell \, \omega_j^{(n)} = \int_{\lambda_L}^{\lambda_U} \lambda^\ell \, d\omega(\lambda), \quad \ell = 0, 1, \ldots, 2n - 1.$$

This means the following:

1. Provided that the problem to be solved is easy, a single numerical indicator (such as the system matrix **A** condition number) can guarantee fast CG convergence. Then, however, one should also ask whether there is a need for the application of a Krylov subspace method such as CG. Most probably, much simpler linear iterative methods would solve the problem at a lower cost.

2. If the problem is difficult, bounds associated with a single numerical indicator do not provide a realistic information on the rate of convergence of Krylov subspace methods.

Hiptmair referred in a quote given in Chapter 4 to difficult problems. The persisting myths about describing convergence of CG by the convergence bounds based on spectral condition number are addressed in the recent paper [71], which builds upon arguments known for a long time. Despite such arguments, the spectral condition number is still widely used as an indicator of the rate of convergence of CG, and a few outlying eigenvalues are considered harmless for *asymptotic* convergence. Even worse, this myth is taught to students all over the world, without taking into account known facts presented in many papers. Focusing on the condition number characterization of the rate of convergence means that after a sophisticated mathematical reasoning in analysis and discretization of an infinite-dimensional PDE problem and after robust and efficient implementations which consider the state of the art parallel computer architectures, the final argument about the rate of convergence of the preconditioned Krylov subspace method is simplified to such an extent, that it can be unrealistic or even wrong.

In particular, it *is not true* that CG (or other Krylov subspace methods used for solving systems of linear algebraic equations with symmetric matrices) applied to a matrix with t distinct well-separated tight clusters of eigenvalues produces in general a large error reduction after t steps; see [124, Sections 5.6.5 and 5.9.1]. This myth was disproved more than 20 years ago; see [82, 182, 85]. Still it is persistently repeated in an authoritative way in the literature. Similarly, it *is not true* that distribution of eigenvalues provides, with no further information on the structure of invariant subspaces, insight into the asymptotic behavior of Krylov subspace methods (such as GMRES) applied to systems with generally nonsymmetric matrices; see [124, Sections 5.7.4, 5.7.6, and 5.11]. As above, the relevant results [86, 84, 4] are (almost) twenty years old.

In Krylov subspace methods with short recurrences, like CG, it must always be taken into account that rounding errors in practical computations mean loss of orthogonality. Apart from very special cases, there is no way to preserve orthogonality among the vectors computed using short recurrences and, at the same time, to compute, using such recurrences, well-conditioned bases of Krylov subspaces. Either we must reorthogonalize, which means involving *long recurrences*, or orthogonality and consequently also linear independence of the computed basis vectors are inevitably lost; see [124, Section 5.9]. Deriving results on the convergence of CG *assuming exact arithmetic* and then applying such results to practical computations without a thorough numerical stability analysis, which would provide an appropriate justification, does not represent an approach conforming with standard mathematical methodology, which relies upon rigor of arguments. Still, this approach is widespread in literature, with possible delays of convergence and limits for the maximal attainable accuracy simply ignored. It should be emphasized that even if local rounding errors at individual arithmetic operations or even at individual CG iterations are small, their effect on the whole solution process can be devastating. The total error is not formed by an *accumulation* of elementary rounding errors[20]. The total error can become large due to the substantial *amplification* of elementary rounding errors; see, e.g., [184, 183, 137, 77, 135] and [124, Sections 5.8–5.10] and the references given there.

When the algebraic problems are nonsymmetric, any convergence analysis based on spectral information becomes not only technically involved but also mathematically rather intricate. It is well known, with the first paper on this subject published in 1994 [86], that for nonsymmetric problems in general *any convergence* of Krylov subspace methods can be observed with matrices having *any eigenvalues*; see [124, Section 5.7] and, for the recent developments, e.g., [53, 136]. This *does not* mean that the convergence of Krylov subspace methods in the nonsymmetric case does not depend on the spectrum of the matrix (or the underlying operator), but that the *spectral information alone* is not sufficient for making *any conclusions* about convergence. Still, despite these well-known results, many papers about preconditioning by deflation and related techniques argue in exactly the opposite way without even mentioning that the structure of invariant subspaces must be taken into account. References to the works which *proved* that information about eigenvalues alone is not sufficient for describing convergence are very often missing.

It might seem that the difficulty with finite-dimensional algebraic descriptions of convergence can be resolved by analyzing convergence of iterative methods in terms of the original infinite-dimensional operators in Banach or Hilbert spaces. This can indeed be done and it can bring very interesting results and insights; see, e.g., [115, 24, 25, 143]. On the other hand, such approaches do not address (and do not answer) questions on the cost of particular finite-dimensional algebraic computations for the given discretized operator and the given discretized right-hand side of the operator equation. Recently Liesen directed us to a visionary paper by Vlastimil Pták titled "What should be a rate of convergence?"; see [157, 123]. In this paper Pták poses requirements for a method of estimating the convergence of iterative processes:

> *1°) It should relate quantities which may be measured or estimated during the actual process.*

> *2°) It should describe accurately in particular the initial stage of the process, not only its asymptotic behavior since, after all, we are interested in keeping the number of steps necessary to obtain a good estimate as low as possible.*

These points, presented by a prominent functional analyst, offer a crystal-clear view. Description of the rate of convergence must serve the main purpose which is *estimating the cost of practical computations*. This cannot be done by any single number-based bounds nor by infinite-dimensional operator polynomial-based bounds. As mentioned in Section 5.2, a priori convergence analysis of Krylov subspace methods should, in our opinion, consider as the base the matching moments model reduction property, which can lead to some new results. Evaluation of the cost of actual computations requires *a posteriori error estimators* which will allow one to take into account all sources of errors.

---

[20]The view that rounding errors can be dangerous because they accumulate is another widespread inaccuracy which distorts the common knowledge of practitioners on the effects of rounding errors. The *amplification* as the main factor has been clearly identified in the early works on numerical stability of the Gaussian elimination; see, e.g., [191].

Chapter 12

Inexact Computations, A Posteriori Error Analysis, and Stopping Criteria

As mentioned in the previous chapters, algebraic computations are not performed exactly. This fact affects the computational cost because a nonnegligible extra effort is needed, in comparison to hypothetical exact computations, in order to achieve the desired accuracy. This can even affect solvability of the problem, because rounding errors may not allow one to get a sufficiently accurate solution. Dealing with inexactness of algebraic computations represents a standard in fields such as nonlinear optimization and also in PDE constrained optimization and domain decomposition techniques; see, e.g., [80, 112, 129, 64, 109]. The same is not true, despite the effort of many authors, in the general numerical PDE context, which is somewhat surprising. As indicated by the quote by Hackbusch given in the introduction, the point is that considering "exact algebraic computations" in solving difficult PDE problems is not only unrealistic but simply a false way of approaching the problem. Accuracy of algebraic computations should be in an appropriate balance with other errors (modeling, linearization, discretization). Setting stringent requirement on accuracy of algebraic computations can inflate the cost of the entire solution process. In summary:

> Algebraic computations are not performed exactly. Algebraic errors, arising from incomplete algebraic computations, from rounding errors, or from both, are generally not negligible, and they must be taken into account in the overall quantification of the error.

The actual error in the solution process is evaluated a posteriori, using estimators which should be fully computable. Ideally, a posteriori error estimators should be:

- reliable, i.e., they should provide guaranteed upper bounds;

- locally efficient, i.e., they should not locally significantly underestimate the true error;

- asymptotically exact, i.e., the ratio of the estimator value and the size of the true error should approach one as the discretization is refined.

The a posteriori error estimates should provide information on the spatial distribution of the error over the domain, including the indication of the sources of the individual error components (discretization, linearization, algebraic computations). There is a vast amount of literature on a posteriori error analysis in PDEs, and the progress achieved in the past decades is remarkable. One can refer to many monographs (see, e.g., [187, 1, 16, 17, 169]) or surveys (see, e.g., [23, 188, 3, 168, 144, 195]). Starting from the early works (see, e.g., [15]), the area has developed in many directions, some of which include errors of algebraic computations; see, e.g., the introductory paper [167], the cascadic CG [50], the FOSLS approach [26], combination of wavelet discretization and iterative methods [38, 39], considering inexact algebraic computations in FEM adaptivity [19, 22, 20, 21, 95], using a posteriori bounds for stopping iterative computations [148, 153, 101, 176, 61], and the references given in [3, Section 4]. Reaching the goal as stated above is challenging, however, and there is still substantial work to be done.

Despite the examples referred to above and the increasing effort of many researchers, a posteriori error analysis in numerical solution of PDEs is still focused on the discretization error, i.e., *the assumption that the algebraic computations are performed exactly* still prevails in the literature; see, e.g., [3, Sections 3 and 4], [187]. The roots of this paradigm can perhaps be linked to using the residual a posteriori estimators and to direct methods for solving linear algebraic systems. (Assuming numerically solvable problems, appropriate methods and implementations, direct methods provide residuals proportional to machine precision unit; see, e.g., [97].) It should be noticed, however, that small algebraic residuals do not necessarily guarantee a small algebraic error; see Chapters 9 and 10.

In this book we focus on iterative methods and, in particular, on CG, and therefore our considerations will continue within this framework. Consider the Galerkin discretization; see Chapters 7 and 8. The quantitative relation between the total error $u - u_h^n$ and the discretization error $u - u_h$, measured in the energy norm $\|\cdot\|_a$, where u_h^n is the approximation to u_h computed at the nth step of CG (considering actual computations *including roundoff*), can be summarized as

$$\|u - u_h^n\|_a^2 = \|u - u_h + u_h - u_h^n\|_a^2 = \|u - u_h\|_a^2 + \|u_h - u_h^n\|_a^2,$$

where $u_h - u_h^n \in V_h$, and therefore the equality follows from the Galerkin orthogonality (7.6). Since

$$u_h = \Phi \mathbf{x}, \quad u_h^n = \Phi \mathbf{x}_n,$$

we immediately get the relationship between the discretization error $\|u - u_h\|_a$ and the algebraic error $\|u_h - u_h^n\|_a = \|\mathbf{x} - \mathbf{x}_n\|_\mathbf{A}$ (see Chapter 6) in the form

$$\|u - u_h^n\|_a^2 = \|u - u_h\|_a^2 + \|\mathbf{x} - \mathbf{x}_n\|_\mathbf{A}^2. \tag{12.1}$$

This separates the contribution of the discretization and the inexact algebraic computation to the size of the total error. The crucial point is that such a separation can unfortunately not be fully achieved within the a posteriori error evaluation.

The difficulty is not only in estimating the algebraic error $\|\mathbf{x} - \mathbf{x}_n\|_\mathbf{A}$. Although no guaranteed tight upper bound can formally be proved due to the presence of rounding errors (see [78]), the techniques which have been developed over more than the past three decades provide reasonably accurate estimates for $\|\mathbf{x} - \mathbf{x}_n\|_\mathbf{A}$ in many practical cases; see [184, 137, 138] and the summary in [124, Sections 5.6 and 5.9]. The difficulty is present also in evaluating the discretization error $u - u_h$, i.e., with estimating $\|u - u_h\|_a$. Indeed, in estimating $\|u - u_h\|_a$ we cannot simply *assume that the algebraic computations are performed exactly*, i.e., simply assume the Galerkin orthogonality (7.6) in the derivation of the estimators; see [3, Section 3], [187]. In determining the values of the estimators, the actually computed quantities (such as residuals and fluxes) are used. For the computed $u - u_h^n$ the Galerkin orthogonality (7.6) does not hold. If the estimators were derived assuming the Galerkin orthogonality, then evaluation of the estimators using the computed quantities violates the assumptions under which they have been derived. Coupling the discretization error and the algebraic error therefore cannot be based on a straightforward application of (12.1) or on similar equations containing the discretization error $u - u_h$.

The difficulty of coupling the discretization error and the algebraic error is nicely illustrated by the following quotes from the works of Roache and Pelletier [170, 171, 151] and Knupp and Salari [110] on *verification* of complex codes for solving challenging practical problems:

- *Most importantly, incomplete iterative convergence will corrupt Verification.* [170, p. 36]

- *It is important that iterative convergence criteria be stringent so as not to confuse the incomplete iteration (residual) error with discretization error.* [170, p. 72]

- *The basic idea, of course, is to make the iteration residual negligible with respect to the discretization error.* [151, p. 424]

- *IICE [incomplete iterative convergence error] can be decreased to the level of roundoff error by using sufficiently tight stopping criteria ... By using tight iterative tolerances and well-conditioned test problems, we can ensure that the numerical error is nearly the same as the discretization error.* [110, p. 15]

- *There may be incomplete iterative convergence (IICE) or round-off-error that is polluting the results. If the code uses an iterative solver, then one must be sure that the iterative stopping criteria is sufficiently tight so that the numerical and discrete solutions are close to one another. Usually in order-verification tests, one sets the iterative stopping criterion to just above the level of machine precision to circumvent this possibility.* [110, p. 31]

What is a natural and standard in verification methodology cannot be, however, applied in practical computations; see also [124, Section 1.1].

Coupling discretization with algebraic solvers while considering inaccurate algebraic computations has been an area of study for many years, with the cascadic CG of Deuflhard [50] and the dual weighted residual methods (DWR) of Rannacher, Becker, and their collaborators [167, 23, 168] as well as other methods of a posteriori error analysis based on duality arguments (see the paper by Giles and Süli [73]) among the pioneering examples; see the survey [3, Section 4]. The existing approaches vary in complying with the requirements of reliability, local efficiency, and asymptotic exactness, and they also differ in using various heuristic arguments and their justifications.

As pointed out in [124, Sections 5.1 and 5.9.4] and [147], a posteriori error analysis should take into account the fact that the spatial distribution of the discretization error and the algebraic error can be very different (see the discussion in Chapter 10). The globally measured energy norm of the error can be misleading in the sense that the algebraic error can dominate the total error in some parts of the domain even if in the globally measured energy norm the algebraic error is much smaller than the energy norm of the discretization error. Within the models with elliptic self-adjoint bounded and α-coercive operators the energy norm is linked to the physics of the problem, and therefore it represents the relevant norm. Still, even if $\|u - u_h\|_a$ and $\|\mathbf{x} - \mathbf{x}_n\|_\mathbf{A}$ were computable (which is a hypothetical assumption which cannot be satisfied in practice), the balance between the discretization error and the algebraic error cannot be based on the equality (12.1).

Discretization mesh and/or polynomial basis functions *adaptivity* is among the paradigms which enable efficient computational solution of challenging problems. The assumption of exact algebraic computations questioned above is also used in the standard proofs of convergence of adaptive finite element methods (AFEM), with the conclusion that (see, e.g., [139])

Any prescribed error tolerance is thus achieved in a finite number of steps.

Here the "step" means a single loop of AFEM consisting of solving the discretized problem, estimating the discretization error, identifying the mesh elements, and refining the discretization. A relevant question is to which extent such result can be rigorously extended to practical computations.

First, mesh refinement can lead to algebraic systems where the maximal attainable accuracy of iterative (or direct) solvers will become important. It may eventually limit the overall accuracy of solving the PDE problem; see [124, Section 5.9.3] and the references given there. In the field of iterative methods one can still encounter the myth that *presumable self-correcting effects* of iteration schemes ensure that an arbitrary accuracy can be reached in iterative computations. This myth is rooted in some form of "self-correcting" effects of some linear iterations; see the description of the related historical development in [124, Section 5.8]. Even then, self-correcting effects do not mean that an arbitrary accuracy can be reached using standard floating point arithmetic (with limited computer arithmetic accuracy). Moreover, the concept of self-correction is in principle not valid in Krylov subspace methods, which are based on projections and matching moments model reduction.

Second, even assuming that the maximal attainable accuracy of the algebraic solvers is not an issue because the iterations will be stopped much before such limiting accuracy is reached, the extension of proofs of AFEM convergence for inexact algebraic computations is rather involved. This can be documented by a beautiful paper by Stevenson [178]. The main result, formulated in Theorem 6.3 of [178], holds under assumptions which are not easy to interpret, although their specification in the paper is done with full rigor. They also refer to the paper [196] for an essential ingredient on the uniform convergence of the multigrid V-cycle on adaptively refined finite element meshes. The analysis of the multigrid algorithm in [196] assumes exact arithmetic, however, and in particular, it assumes an *exact inversion of the operator on the coarsest mesh*. In practical computations such assumptions are not satisfied. The effects of rounding errors in multilevel computations are yet to be investigated.

Chapter 13

Summary and Outlook

The road from PDEs through functional analysis and discretization to iterative methods is long and not without danger. Remarkable achievements can be observed along all its parts. Difficulty is hidden in putting together the results achieved within the individual parts in order to achieve maximal efficiency in using the resources available. In order to achieve this, builders of the individual parts should be in close mutual interaction. Deep expertise within one part is not sufficient and cannot compensate for deficiencies at other parts.

In the motto above we used the phrase from the Tolkien fantasy book about the journey of the hobbit Bilbo "there and back again." We believe that a one-way passage from PDEs through functional analysis to iterative methods is not enough and that one must go there and back again (perhaps several times) in order to understand the solution process as a whole. When the functional equation $\mathscr{A}u = b$ is transformed by operator preconditioning, based on the choice of the inner product and the associated Riesz map, into the equation $\tau\mathscr{A}u = \tau b$, the crucial question is, what is the optimal preconditioning, where *optimality* must consider the whole solution process. Such optimality cannot be reduced to evaluations using the idea of conditioning. As Forsythe pointed out in 1953 in his article published in the Bulletin of the American Mathematical Society [67, p. 318]:

> *There is a great need for clarification of the group of ideas associated with "condition." With the concept of "ill-conditioned" systems $Ax = b$ goes the idea of "preconditioning" them.*

Here we do not argue against using condition numbers and the bounds on them *where appropriate*. We argue against using them as general unquestioned tools which are considered fully descriptive and which are used as arguments closing the door for further investigation.

This is related to another point. When Krylov subspace methods are considered, the question of approximating the *operators* $(\tau\mathscr{A})^{-1}$ or $(\mathscr{I} - \tau\mathscr{A})^{-1}$, regardless how useful it might be in considering some particular issues, must be distinguished from the question of approximating the solution $(\tau\mathscr{A})^{-1}b$ using the sequence of vectors $\tau b, (\tau\mathscr{A})\tau b, (\tau\mathscr{A})^2\tau b, \ldots$ Looking even farther ahead, the efficiency of computing numerical approximations to the solution of the finite-dimensional algebraic system $\mathbf{A}\mathbf{x} = \mathbf{b}$ cannot in general be evaluated using the results on the approximation of infinite-dimensional operators $(\tau \mathscr{A})^{-1}$. Therefore, in order to consider the optimality of operator preconditioning, one must see the *whole journey* up to the finite-dimensional algebraic computation with *including the implementation and numerical stability issues.*

As an important point, in particular in dealing with subjects across many disciplines, justification of computational approaches using simple examples cannot be valid without considering issues which need not be apparent on such simple examples. Model problems serve as excellent tools for developing ideas and for examining their relevance. In particular, if a difficulty can be observed using the simplest possible model problem, such as the difference between spatial distribution of discretization and algebraic errors (see [147]), then it indicates that such difficulty should be taken seriously in considering practical computations. Model problems cannot be used, however, with the same relevance for the evaluation and comparison of various approaches intended for solving challenging problems.

Model problems used throughout the literature are often too simple to observe difficulties which can arise in real computations. This concerns analytical properties of models with various consequences for assumptions which different approaches might be based on, as well as the evaluation of computational efficiency. Here, behavior of CG gives an instructive example. For the discretized (close-to-) Laplace operator, CG need not exhibit fast convergence and it is meaningless to use such model problems for comparison of CG with other methods such as multigrid. A different confusion may arise from the fact that loss of orthogonality in finite precision CG computations is for such model problems typically not an issue one is concerned with; see [124, Theorem 3.4.12 and Sections 5.9.1 and 5.9.4, in particular p. 325 and Figure 5.29] for a detailed explanation. This certainly does not mean that rounding errors do not significantly affect CG computations in solving practical problems. Just the opposite takes place. Significant role of rounding errors can, moreover, be demonstrated on well-conditioned diagonal matrices of very small size (see, e.g., [182], [85]). It is well understood and *theoretically justified* that such elementary models capture the essence of the general phenomenon, which depends on the associated Riemann–Stieltjes disctribution function; see Section 5.2.

Within the abstract setting adopted throughout our book, PDE problems are described in the form of the functional equation

$$\mathscr{A} x = b, \quad \mathscr{A} : V \to V^{\#}, \quad x \in V, \quad b \in V^{\#},$$

where the linear operator $\mathscr{A}$ is self-adjoint (with respect to the duality pairing $\langle \cdot, \cdot \rangle$), bounded, and coercive. This naturally leads to preconditioned conjugate gradients (PCG) for solving the associated algebraic problem. The setting presented is certainly not restrictive and can be generalized to functional equations with other properties of $\mathscr{A}$ and other algebraic solvers.

When algebraic preconditioning is motivated by operator preconditioning, here described via the Riesz map τ determined by the choice of the inner product $(\cdot, \cdot)$, we have illustrated the solution process by the scheme

$$\{\mathscr{A}, b, \tau\} \to \{\mathscr{A}_h, b_h, \tau\} \to \{\mathbf{A}_h, \mathbf{b}_h, \mathbf{M}_h\} \to \text{PCG applied to } \mathbf{A}_h \mathbf{x}_h = \mathbf{b}_h.$$

The discretization step from $\mathscr{A} x = b$ to $\mathscr{A}_h x_h = b_h$ is the subject of highly developed mathematical technology (with FEM as the prevailing paradigm) which aims at ensuring convergence of x_h to x as the discretization parameter h approaches zero. At the same time it should enable efficient numerical approximation of the solution $\mathbf{x}_h$ of the algebraic problem. A principal issue, which makes the discretization step and also its coupling with algebraic computations difficult, is in the fact that $\mathscr{A}_h x_h = b_h$ cannot be considered as an approximation of $\mathscr{A}x = b$ in the sense that $\mathscr{A}_h$ converges in a suitable norm to $\mathscr{A}$. If V is infinite-dimensional, then the invertible operator $\mathscr{A}$ is not compact and therefore cannot be given as a limit of finite-dimensional (compact) operators. Related to that, behavior of PCG applied to the algebraic equation $\mathbf{A}_h \mathbf{x}_h = \mathbf{b}_h$ cannot be, in general, easily described on the functional analytic level using the operator $\mathscr{A}$. This point is even more important for Krylov subspace methods used with algebraic problems arising from non self-adjoint operators.

The common approach to numerical solution of PDEs can also be illustrated by the following scheme which starts with the bilinear form representation of the PDE (weak formulation) and does not consider the interplay of the inner product $(\cdot, \cdot)$ with the duality pairing $\langle \cdot, \cdot \rangle$ defining the Riesz map,

$$a(\cdot, \cdot), \langle b, \cdot \rangle \rightarrow \mathbf{A}_h, \mathbf{b}_h \rightarrow \text{ algebraic preconditioner } \rightarrow \text{ PCG applied to } \mathbf{A}_h \mathbf{x}_h = \mathbf{b}_h.$$

Many times this process is broken into pieces. As an example, discretization is completely separated from the construction of preconditioners, and a posteriori analysis of the discretization errors is still many times separated from inexact algebraic computations and evaluation of the algebraic error. (This contradicts the basic principle of using iterative methods where the main advantage is stopping the iteration process whenever the appropriate accuracy has been reached.)

One of the main points made within our book is that it is convenient to consider discretization and preconditioning closely linked together. Discretization which does not sufficiently respect the nature of the infinite-dimensional problem can lead to finite-dimensional algebraic problems which are difficult to precondition or which require very intriguing preconditioning. As suggested by several techniques used throughout decades, preconditioning can always be linked with transformation of the discretization basis. The simple and general transformation formulas presented in Chapter 8 open a possible way for further research, which can start from interpreting algebraically constructed highly efficient preconditioners as the corresponding discretization basis (and the associated inner product) transformations and from investigating whether similar bases can be used directly for the discretization of the analogous infinite-dimensional problems.

Without diminishing in the slightest the principal role of the discretization step, mathematically it provides a sophisticated technology which enables the approximation of an infinite-dimensional operator $\mathscr{A}$ by suitable finite-dimensional counterparts. In both schemes presented above in this summary, PCG is applied to the finite-dimensional algebraic system. Assume, for the moment, that we are able to perform operations with $\mathscr{A}$ and τ, i.e., that we are able to construct Krylov subspaces

$$K_n(\tau \mathscr{A}, \tau r_0) = \text{span}\{\tau r_0, \tau \mathscr{A} \tau r_0, \ldots, (\tau \mathscr{A})^{n-1} \tau r_0\}, \qquad n = 1, 2, \ldots,$$

where $r_0 = b - \mathscr{A}x_0$, x_0 is an initial approximation to the solution $x \in V$. Then n steps of CG applied to $\mathscr{A}x = b$ provide without any intermediate discretization the n-dimensional algebraic system with the Jacobi matrix $\mathbf{T}_n$ (see Chapter 5):

$$\mathbf{T}_n \mathbf{y}_n = \|\tau r_0\|_V \mathbf{e}_1, \qquad x_n = x_0 + Q_n \mathbf{y}_n,$$

where $Q_n = (q_1, \ldots, q_n)$ represents the orthonormal basis of the finite-dimensional Krylov subspace $K_n \subset V$ constructed via the infinite-dimensional Lanczos process starting with τr_0. (The Jacobi matrix $\mathbf{T}_n$ and the orthonormal basis Q_n are not explicitly computed in the infinite-dimensional CG described in Chapter 5.) Here the discretization orthogonally projects the original operator $\mathscr{A}$ onto K_n and the error analysis can be viewed as evaluation of the error in the associated Gauss–Christoffel quadrature.

The assumption that we are able to *compute* $K_n \subset V$ is certainly too restrictive. Nevertheless, we believe that it is worth investigating this *model reduction approach*, assuming that the operations with $\mathcal{A}$ and τ can be suitably approximated. If feasible, it may lead to interesting alternatives to the state-of-the-art approaches. Whether there is a chance to develop techniques applicable to important practical problems is a question for further research. We believe that it is worth trying.

It is fair to admit that much more could be added or, even more strongly, should be added. The whole setting used in the text is of limited nature regarding practical applications. This setting has been used because it allows for clear mathematical structure which can be presented in a concise way and which is the basis for the whole text. We believe that an analogous approach could be used with appropriate extensions also for more complicated problems.

It would certainly be very convincing to present examples of practical computations in solving large-scale industrial problems, which would demonstrate the potential of the ideas and philosophy presented in this book. As Cornelius Lanczos wrote in his essay [119] quoted in the preface:

> *A new idea in science will be of constructive value only if the edifice is strong enough to withstand all the objections that can be raised against it. This is a criterion which is frequently not fulfilled and thus many otherwise excellent ideas perish by the wayside. It is not enough to satisfy the sympathisers. One must also be able to silence the voice of the* advocatus diaboli.
>
> *Periods of great leaps forward and periods of consolidation follow thus in ever repeating cycles. Relatively stable periods alternate with sudden bursts of feverish activity in which old values are abandoned and when it looks as if everything must be built up again from the foundation, although in fact extensive parts of the building are saved and integrated with the new scheme in a new interpretation.*

It is apparent from the numerous quotes that most of the ideas presented in our book were presented before in some form by many distinguished scientists. We have tried to put them within a unifying concept, and we have perhaps interpreted some of them in a new way or have emphasized their mutual interconnections and importance.

As has been pointed out, many authors have successfully used related ideas in different settings for solving practical problems. As for an efficient use of the unifying strategies outlined in our text for practical industrial problems, we must admit that we are not there yet. We will continue within our abilities to work in that direction. We also would like to present this thought-provoking text as a possible invitation to all readers who might become interested in a similar effort. We would be grateful to anyone for sharing with us their experience and the results obtained.

Bibliography

[1] M. AINSWORTH AND J. T. ODEN, *A Posteriori Error Estimation in Finite Element Analysis*, Pure and Applied Mathematics, Wiley-Interscience [John Wiley & Sons], New York, 2000. (Cited on p. 78)

[2] B. AKSOYLU, S. BOND, AND M. HOLST, *An odyssey into local refinement and multilevel preconditioning. III. Implementation and numerical experiments*, SIAM J. Sci. Comput., 25 (2003), pp. 478–498. (Cited on p. 59)

[3] M. ARIOLI, J. LIESEN, A. MIĘDLAR, AND Z. STRAKOŠ, *Interplay between discretization and algebraic computation in adaptive numerical solution of elliptic PDE problems*, GAMM-Mitt., 36 (2013), pp. 102–129. (Cited on pp. 39, 69, 78, 79)

[4] M. ARIOLI, V. PTÁK, AND Z. STRAKOŠ, *Krylov sequences of maximal length and convergence of GMRES*, BIT, 38 (1998), pp. 636–643. (Cited on p. 74)

[5] D. N. ARNOLD, *Mixed finite element methods for elliptic problems*, Comput. Methods Appl. Mech. Engrg., 82 (1990), pp. 281–300. Reliability in computational mechanics (Austin, TX, 1989). (Cited on p. 20)

[6] D. N. ARNOLD, *Stability, consistency, and convergence of numerical discretizations*, in Encyclopedia of Applied and Computational Mathematics, B. Engquist, ed., Springer, 2015. (Cited on pp. 19, 63)

[7] D. N. ARNOLD, R. S. FALK, AND R. WINTHER, *Preconditioning discrete approximations of the Reissner-Mindlin plate model*, RAIRO Modél. Math. Anal. Numér., 31 (1997), pp. 517–557. (Cited on p. 30)

[8] ——, *Preconditioning in H(div) and applications*, Math. Comp., 66 (1997), pp. 957–984. (Cited on p. 30)

[9] ——, *Finite element exterior calculus: from Hodge theory to numerical stability*, Bull. Amer. Math. Soc. (N.S.), 47 (2010), pp. 281–354. (Cited on pp. 19, 63, 65, 70)

[10] K. ATKINSON AND W. HAN, *Theoretical Numerical Analysis: A Functional Analysis Framework*, vol. 39 of Texts in Applied Mathematics, Springer, Dordrecht, third ed., 2009. (Cited on p. 63)

[11] O. AXELSSON AND V. A. BARKER, *Finite Element Solution of Boundary Value Problems: Theory and Computation*, vol. 35 of Classics in Applied Mathematics, SIAM, Philadelphia, PA, 2001. Reprint of the 1984 original. (Cited on pp. 55, 58)

[12] O. AXELSSON AND J. KARÁTSON, *Equivalent operator preconditioning for elliptic problems*, Numer. Algorithms, 50 (2009), pp. 297–380. (Cited on pp. 30, 31)

[13] O. AXELSSON AND P. S. VASSILEVSKI, *Algebraic multilevel preconditioning methods. I*, Numer. Math., 56 (1989), pp. 157–177. (Cited on p. 59)

[14] ——, *A survey of multilevel preconditioned iterative methods*, BIT, 29 (1989), pp. 769–793. (Cited on p. 59)

[15] I. BABUŠKA AND W. C. RHEINBOLDT, *Error estimates for adaptive finite element computations*, SIAM J. Numer. Anal., 15 (1978), pp. 736–754. (Cited on p. 78)

[16] I. BABUŠKA AND T. STROUBOULIS, *The Finite Element Method and Its Reliability*, Numerical Mathematics and Scientific Computation, The Clarendon Press Oxford University Press, New York, 2001. (Cited on pp. 69, 78)

[17] W. BANGERTH AND R. RANNACHER, *Adaptive Finite Element Methods for Differential Equations*, Lectures in Mathematics ETH Zürich, Birkhäuser Verlag, Basel, 2003. (Cited on p. 78)

[18] R. BECKER, C. JOHNSON, AND R. RANNACHER, *Adaptive error control for multigrid finite element methods*, Computing, 55 (1995), pp. 271–288. (Cited on pp. 69, 73)

[19] R. BECKER AND S. MAO, *An optimally convergent adaptive mixed finite element method*, Numer. Math., 111 (2008), pp. 35–54. (Cited on p. 78)

[20] ——, *Convergence and quasi-optimal complexity of a simple adaptive finite element method*, M2AN Math. Model. Numer. Anal., 43 (2009), pp. 1203–1219. (Cited on p. 78)

[21] R. BECKER, S. MAO, AND Z.-C. SHI, *A convergent nonconforming adaptive finite element method with quasi-optimal complexity*, SIAM J. Numer. Anal., 47 (2010), pp. 4639–4659. (Cited on p. 78)

[22] ——, *A convergent adaptive finite element method with optimal complexity*, Electron. Trans. Numer. Anal., 30 (2008), pp. 291–304. (Cited on p. 78)

[23] R. BECKER AND R. RANNACHER, *An optimal control approach to a posteriori error estimation in finite element methods*, Acta Numer., 10 (2001), pp. 1–102. (Cited on pp. 78, 79)

[24] B. BECKERMANN AND A. B. J. KUIJLAARS, *On the sharpness of an asymptotic error estimate for conjugate gradients*, BIT, 41 (2001), pp. 856–867. (Cited on p. 75)

[25] ——, *Superlinear CG convergence for special right-hand sides*, Electron. Trans. Numer. Anal., 14 (2002), pp. 1–19. (Cited on p. 75)

[26] M. BERNDT, T. A. MANTEUFFEL, AND S. F. MCCORMICK, *Local error estimates and adaptive refinement for first-order system least squares (FOSLS)*, Electron. Trans. Numer. Anal., 6 (1997), pp. 35–43. (Cited on pp. 15, 78)

[27] D. BOFFI, F. BREZZI, L. F. DEMKOWICZ, R. G. DURÁN, R. S. FALK, AND M. FORTIN, *Mixed Finite Elements, Compatibility Conditions, and Applications*, vol. 1939 of Lecture Notes in Mathematics, Springer-Verlag, Berlin, 2008. Lectures given at the C.I.M.E. Summer School held in Cetraro, June 26–July 1, 2006, Edited by D. Boffi and L. Gastaldi. (Cited on p. 19)

[28] D. BOFFI, F. BREZZI, AND M. FORTIN, *Mixed Finite Element Methods and Applications*, Springer Series in Computational Mathematics 44. Springer, Berlin, 2013. (Cited on p. 19)

[29] J. BRAMBLE, J. PASCIAK, AND J. XU, *Parallel multilevel preconditioners*, Math. Comp., 55 (1990), pp. 1–22. (Cited on p. 59)

[30] S. C. BRENNER AND L. R. SCOTT, *The Mathematical Theory of Finite Element Methods*, vol. 15 of Texts in Applied Mathematics, Springer-Verlag, New York, second ed., 2002. (Cited on pp. 23, 63, 65)

[31] C. BREZINSKI, *History of Continued Fractions and Padé Approximants*, vol. 12 of Springer Series in Computational Mathematics, Springer-Verlag, Berlin, 1991. (Cited on p. 43)

[32] ——, *The methods of Vorobyev and Lanczos*, Linear Algebra Appl., 234 (1996), pp. 21–41. (Cited on p. 43)

[33] ——, *Projection Methods for Systems of Equations*, vol. 7 of Studies in Computational Mathematics, North-Holland Publishing Co., Amsterdam, 1997. (Cited on p. 43)

[34] H. BREZIS, *Functional Analysis, Sobolev Spaces and Partial Differential Equations*, Universitext, Springer, New York, 2011. (Cited on p. 7)

[35] H. BREZIS AND F. BROWDER, *Partial differential equations in the 20th century*, Adv. Math., 135 (1998), pp. 76–144. (Cited on p. 11)

[36] M. BULÍČEK, E. FEIREISL, AND J. MÁLEK, *A Navier-Stokes-Fourier system for incompressible fluids with temperature dependent material coefficients*, Nonlinear Anal. Real World Appl., 10 (2009), pp. 992–1015. (Cited on p. 14)

[37] M. BULÍČEK, P. GWIAZDA, J. MÁLEK, AND A. ŚWIERCZEWSKA-GWIAZDA, *On unsteady flows of implicitly constituted incompressible fluids*, SIAM J. Math. Anal., 44 (2012), pp. 2756–2801. (Cited on p. 19)

[38] C. BURSTEDDE AND A. KUNOTH, *Fast iterative solution of elliptic control problems in wavelet discretization*, J. Comput. Appl. Math., 196 (2006), pp. 299–319. (Cited on p. 78)

[39] ——, *A wavelet-based nested iteration-inexact conjugate gradient algorithm for adaptively solving elliptic PDEs*, Numer. Algorithms, 48 (2008), pp. 161–188. (Cited on pp. 70, 78)

[40] Z. CAI, R. LAZAROV, T. A. MANTEUFFEL, AND S. F. McCORMICK, *First-order system least squares for second-order partial differential equations. Part I*, SIAM J. Numer. Anal., 31 (1994), pp. 1785–1799. (Cited on p. 19)

[41] Z. CAI, T. A. MANTEUFFEL, S. F. McCORMICK, AND J. RUGE, *First-order system $\mathscr{L}\mathscr{L}^*$ (FOSLL*): scalar elliptic partial differential equations*, SIAM J. Numer. Anal., 39 (2001), pp. 1418–1445. (Cited on p. 19)

[42] S. H. CHRISTIANSEN AND J.-C. NÉDÉLEC, *Des préconditionneurs pour la résolution numérique des équations intégrales de frontière de l'acoustique*, C. R. Acad. Sci. Paris Sér. I Math., 330 (2000), pp. 617–622. (Cited on p. 30)

[43] ——, *Des préconditionneurs pour la résolution numérique des équations intégrales de frontière de l'électromagnétisme*, C. R. Acad. Sci. Paris Sér. I Math., 331 (2000), pp. 733–738. (Cited on p. 30)

[44] E. B. CHRISTOFFEL, *Über die Gaußische Quadratur und eine Verallgemeinerung derselben*, J. Reine Angew. Math., 55 (1858), pp. 61–82. Reprinted in Gesammelte mathematische Abhandlungen I (B. G. Teubner, Leipzig, 1910), pp. 65–87. (Cited on p. 45)

[45] P. G. CIARLET, *The Finite Element Method for Elliptic Problems*, vol. 40 of Classics in Applied Mathematics, SIAM, Philadelphia, PA, 2002. Reprint of the 1978 original published by North-Holland, Amsterdam. (Cited on p. 63)

[46] J. B. CONWAY, *A Course in Functional Analysis*, vol. 96 of Graduate Texts in Mathematics, Springer-Verlag, New York, second ed., 1990. (Cited on p. 63)

[47] R. D. COOK, D. S. MALKUS, M. E. PLESHA, AND R. J. WITT, *Concepts and Applications of Finite Element Analysis*, John Wiley & Sons, fourth ed., 2001. (Cited on p. 67)

[48] W. DAHMEN AND A. KUNOTH, *Multilevel preconditioning*, Numer. Math., 63 (1992), pp. 315–344. (Cited on p. 59)

[49] J. W. DANIEL, *Convergence of the conjugate gradient method with computationally convenient modifications*, Numer. Math., 10 (1967), pp. 125–131. (Cited on p. 38)

[50] P. DEUFLHARD, *Cascadic conjugate gradient methods for elliptic partial differential equations: algorithm and numerical results*, in Domain Decomposition Methods in Scientific and Engineering Computing (University Park, PA, 1993), vol. 180 of Contemp. Math., Amer. Math. Soc., Providence, RI, 1994, pp. 29–42. (Cited on pp. 78, 79)

[51] E. DIBENEDETTO, *Partial Differential Equations*, Cornerstones, Birkhäuser Boston Inc., Boston, MA, second ed., 2010. (Cited on p. 7)

[52] O. DUBOIS, M. J. GANDER, S. LOISEL, A. ST-CYR, AND D. B. SZYLD, *The optimized Schwarz method with a coarse grid correction*, SIAM J. Sci. Comput., 34 (2012), pp. A421–A458. (Cited on p. 70)

[53] J. DUINTJER TEBBENS AND G. MEURANT, *Any Ritz value behavior is possible for Arnoldi and for GMRES*, SIAM J. Matrix Anal. Appl., 33 (2012), pp. 958–978. (Cited on p. 75)

[54] N. DUNFORD AND J. T. SCHWARTZ, *Linear Operators. I. General Theory*, With the assistance of W. G. Bade and R. G. Bartle. Pure and Applied Mathematics, Vol. 7, Interscience Publishers, Inc., New York, 1958. (Cited on p. 63)

[55] E. G. D'YAKONOV, *On an iterative method for the solution of a system of finite-difference equations*, Dokl. Akad. Nauk, 138 (1961), pp. 522–526. (Cited on p. 30)

[56] ——, *The construction of iterative methods based on the use of spectrally equivalent operators*, U.S.S.R Comput. Math. and Math. Phys., 6 (1966), pp. 14–46. (Cited on p. 30)

[57] I. EKELAND AND R. TÉMAM, *Convex Analysis and Variational Problems*, vol. 28 of Classics in Applied Mathematics, SIAM, Philadelphia, PA, English ed., 1999. Translated from the French. (Cited on p. 20)

[58] H. C. ELMAN, A. RAMAGE, AND D. J. SILVESTER, *IFISS: A computational laboratory for investigating incompressible flow problems*, Technical Report MIMS EPrint: 2012.81, Manchester Institute for Mathematical Sciences, School of Mathematics, The University of Manchester, 2012. (Cited on p. 30)

[59] H. C. ELMAN, D. J. SILVESTER, AND A. J. WATHEN, *Finite Elements and Fast Iterative Solvers: With Applications in Incompressible Fluid Dynamics*, Numerical Mathematics and Scientific Computation, Oxford University Press, New York, second ed., 2014. (Cited on p. 30)

[60] M. ENGELI, TH. GINSBURG, H. RUTISHAUSER, AND E. STIEFEL, *Refined iterative methods for computation of the solution and the eigenvalues of self-adjoint boundary value problems*, Mitt. Inst. Angew. Math. Zürich, 8 (1959). (Cited on p. 15)

[61] A. ERN AND M. VOHRALÍK, *Adaptive inexact Newton methods with a posteriori stopping criteria for nonlinear diffusion PDEs*, SIAM J. Sci. Comput., 35 (2013), pp. A1761–A1791. (Cited on p. 78)

[62] L. C. EVANS, *Partial Differential Equations*, vol. 19 of Graduate A Studies in Mathematics, American Mathematical Society, Providence, RI, second ed., 2010. (Cited on pp. 7, 13, 15, 17, 19, 23, 27)

[63] V. FABER, T. A. MANTEUFFEL, AND S. V. PARTER, *On the theory of equivalent operators and application to the numerical solution of uniformly elliptic partial differential equations*, Adv. in Appl. Math., 11 (1990), pp. 109–163. (Cited on pp. 30, 32, 50)

[64] CH. FARHAT, J. MANDEL, AND F.-X. ROUX, *Optimal convergence properties of the FETI domain decomposition method*, Comput. Methods Appl. Mech. Engrg., 115 (1994), pp. 365–385. (Cited on pp. 70, 77)

[65] CH. FARHAT AND F.-X. ROUX, *A method of finite element tearing and interconnecting and its parallel solution algorithm*, Int. J. Numer. Meth. Engrg., 32 (1991), pp. 1205–1227. (Cited on p. 70)

[66] E. FEIREISL, *Dynamics of Viscous Compressible Fluids*, vol. 26 of Oxford Lecture Series in Mathematics and its Applications, Oxford University Press, Oxford, 2004. (Cited on p. 14)

[67] G. E. FORSYTHE, *Solving linear algebraic equations can be interesting*, Bull. Amer. Math. Soc., 59 (1953), pp. 299–329. (Cited on p. 81)

[68] A. D. FREED AND D. R. EINSTEIN, *An implicit elastic theory for lung parenchyma*, Int. J. Eng. Sci., 62 (2013), pp. 31–47. (Cited on p. 19)

[69] C. F. GAUSS, *Methodus nova integralium valores per approximationem inveniendi*, Commentationes Societatis Regiae Scientarium Gottingensis, (1814), pp. 39–76. Reprinted in Werke, Band III (Göttingen, 1876), pp. 163–196. (Cited on p. 45)

[70] W. GAUTSCHI, *A survey of Gauss-Christoffel quadrature formulae*, in E. B. Christoffel (Aachen/Monschau, 1979), Birkhäuser, Basel, 1981, pp. 72–147. (Cited on p. 43)

[71] T. GERGELITS AND Z. STRAKOŠ, *Composite convergence bounds based on Chebyshev polynomials and finite precision conjugate gradient computations*, Numer. Algorithms, 65 (2014), pp. 759–782. (Cited on p. 74)

[72] D. GILBARG AND N. S. TRUDINGER, *Elliptic Partial Differential Equations of Second Order*, Classics in Mathematics, Springer-Verlag, Berlin, 2001. Reprint of the 1998 edition. (Cited on p. 7)

[73] M. B. GILES AND E. SÜLI, *Adjoint methods for PDEs: a posteriori error analysis and postprocessing by duality*, Acta Numer., 11 (2002), pp. 145–236. (Cited on pp. 69, 79)

[74] R. GLOWINSKI, *Finite Element Methods for Incompressible Viscous Flow*, in Handbook of Numerical Analysis, Vol. IX, North-Holland, Amsterdam, 2003. (Cited on p. 32)

[75] M. S. GOCKENBACH, *Understanding and Implementing the Finite Element Method*, SIAM, Philadelphia, PA, 2006. (Cited on pp. 3, 59, 70)

[76] ——, *Partial Differential Equations: Analytical and Numerical Methods*, SIAM, Philadelphia, PA, second ed., 2010. (Cited on pp. 15, 23, 63, 70)

[77] G. H. GOLUB AND G. MEURANT, *Matrices, Moments and Quadrature with Applications*, Princeton Series in Applied Mathematics, Princeton University Press, Princeton, NJ, 2010. (Cited on p. 75)

[78] G. H. GOLUB AND Z. STRAKOŠ, *Estimates in quadratic formulas*, Numer. Algorithms, 8 (1994), pp. 241–268. (Cited on p. 78)

[79] G. H. GOLUB AND J. H. WELSCH, *Calculation of Gauss quadrature rules*, Math. Comp. 23, (1969), pp. A1–A10. (Cited on p. 56)

[80] S. GÖTSCHEL AND M. WEISER, *Lossy compression for PDE-constrained optimization: Adaptive error control*, Comput. Optim. Appl. (accepted for publication), 2014. (Cited on p. 77)

[81] T. GRAUSCHOPF, M. GRIEBEL, AND H. REGLER, *Additive multilevel preconditioners based on bilinear interpolation, matrix-dependent geometric coarsening and algebraic multigrid coarsening for second-order elliptic PDEs*, Appl. Numer. Math., 23 (1997), pp. 63–95. Multilevel methods (Oberwolfach, 1995). (Cited on pp. 59, 70)

[82] A. GREENBAUM, *Behavior of slightly perturbed Lanczos and conjugate-gradient recurrences*, Linear Algebra Appl., 113 (1989), pp. 7–63. (Cited on p. 74)

[83] ——, *Iterative Methods for Solving Linear Systems*, vol. 17 of Frontiers in Applied Mathematics, SIAM, Philadelphia, PA, 1997. (Cited on pp. 55, 58)

[84] A. GREENBAUM, V. PTÁK, AND Z. STRAKOŠ, *Any nonincreasing convergence curve is possible for GMRES*, SIAM J. Matrix Anal. Appl., 17 (1996), pp. 465–469. (Cited on p. 74)

[85] A. GREENBAUM AND Z. STRAKOŠ, *Predicting the behavior of finite precision Lanczos and conjugate gradient computations*, SIAM J. Matrix Anal. Appl., 13 (1992), pp. 121–137. (Cited on pp. 74, 82)

[86] ———, *Matrices that generate the same Krylov residual spaces*, in Recent Advances in Iterative Methods, vol. 60 of IMA Vol. Math. Appl., Springer, New York, 1994, pp. 95–118. (Cited on pp. 74, 75)

[87] M. GRIEBEL AND P. OSWALD, *On the abstract theory of additive and multiplicative Schwarz algorithms*, Numer. Math., 70 (1995), pp. 163–180. (Cited on p. 59)

[88] P. GRISVARD, *Elliptic Problems in Nonsmooth Domains*, vol. 69 of Classics in Applied Mathematics, SIAM, Philadelphia, PA, 2011. (Cited on p. 11)

[89] ———, *Singularities in Boundary Value Problems*, vol. 22 of Recherches en Mathématiques Appliquées [Research in Applied Mathematics], Masson, Paris, 1992. (Cited on p. 11)

[90] J. E. GUNN, *The numerical solution of $\nabla \cdot a\nabla u = f$ by a semi-explicit alternating-direction iterative technique*, Numer. Math., 6 (1964), pp. 181–184. (Cited on p. 30)

[91] ———, *The solution of elliptic difference equations by semi-explicit iterative techniques*, J. Soc. Indust. Appl. Math. Ser. B Numer. Anal., 2 (1965), pp. 24–45. (Cited on p. 30)

[92] A. GÜNNEL, R. HERZOG, AND E. SACHS, *A note on preconditioners and scalar products for Krylov methods in Hilbert space*, Electron. Trans. Numer. Anal., 41 (2014), pp. 13–20. (Cited on pp. 32, 41, 55)

[93] W. HACKBUSCH, *Iterative Solution of Large Sparse Systems of Equations*, vol. 95 of Applied Mathematical Sciences, Springer-Verlag, New York, 1994. Translated and revised from the 1991 German original. (Cited on p. 58)

[94] L. A. HAGEMAN AND D. M. YOUNG, *Applied Iterative Methods*, Academic Press Inc., New York, 1981. (Cited on p. 58)

[95] H. HARBRECHT AND R. SCHNEIDER, *On error estimation in finite element methods without having Galerkin orthogonality*, Preprint 457, Berichtsreihe des SFB 611, Universität Bonn, 2009. (Cited on p. 78)

[96] M. R. HESTENES AND E. STIEFEL, *Methods of conjugate gradients for solving linear systems*, J. Research Nat. Bur. Standards, 49 (1952), pp. 409–436 (1953). (Cited on pp. 15, 55)

[97] N. J. HIGHAM, *Accuracy and Stability of Numerical Algorithms*, SIAM, Philadelphia, PA, second ed., 2002. (Cited on pp. 66, 78)

[98] R. HIPTMAIR, *Operator preconditioning*, Comput. Math. Appl., 52 (2006), pp. 699–706. (Cited on pp. 30, 32, 57)

[99] C. G. J. JACOBI, *Ueber Gauss neue Methode, die Werthe der Integrale näherungsweise zu finden*, J. Reine Angew. Math., 1 (1826), pp. 301–308. Reprinted in Gesammelte Werke, 6. Band (Reimer, Berlin, 1891), pp. 3–11. (Cited on p. 45)

[100] M. JAROŠOVÁ, A. KLAWONN, AND O. RHEINBACH, *Projector preconditioning and transformation of basis in FETI-DP algorithms for contact problems*, Math. Comput. Simul., 82 (2012), pp. 1894–1907. (Cited on pp. 51, 59)

[101] P. JIRÁNEK, Z. STRAKOŠ, AND M. VOHRALÍK, *A posteriori error estimates including algebraic error and stopping criteria for iterative solvers*, SIAM J. Sci. Comput., 32 (2010), pp. 1567–1590. (Cited on pp. 69, 78)

[102] R. KANNAN, S. HENDRY, N. J. HIGHAM, AND F. TISSEUR, *Detecting the causes of ill-conditioning in structural finite element models*, Computers & Structures, 133 (2014), pp. 79–89. (Cited on p. 67)

[103] R. C. KIRBY, *From functional analysis to iterative methods*, SIAM Rev., 52 (2010), pp. 269–293. (Cited on pp. 23, 30, 31)

[104] A. KLAWONN, *An optimal preconditioner for a class of saddle point problems with a penalty term, Part II: General theory*, Technical Report 14/95, Westfälische Wilhelms-Universität Münster, Germany, 04 1995. (Cited on p. 30)

[105] ——, *Preconditioners for indefinite problems*, PhD thesis, Universität Münster, 1996. (Cited on p. 30)

[106] A. KLAWONN, L. F. PAVARINO, AND O. RHEINBACH, *Spectral element FETI-DP and BDDC preconditioners with multi-element subdomains*, Comput. Methods Appl. Mech. Engrg., 198 (2008), pp. 511–523. (Cited on p. 70)

[107] A. KLAWONN AND O. RHEINBACH, *Deflation, projector preconditioning, and balancing in iterative substructuring methods: connections and new results*, SIAM J. Sci. Comput., 34 (2012), pp. A459–A484. (Cited on p. 70)

[108] A. KLAWONN AND G. STARKE, *A preconditioner for the equations of linear elasticity discretized by the PEERS element*, Numer. Linear Algebra Appl., 11 (2004), pp. 493–510. (Cited on p. 30)

[109] A. KLAWONN AND O. B. WIDLUND, *A domain decomposition method with Lagrange multipliers for linear elasticity*, in Eleventh International Conference on Domain Decomposition Methods (London, 1998), DDM.org, Augsburg, 1999, pp. 49–56. (Cited on p. 77)

[110] P. KNUPP AND K. SALARI, *Verification of Computer Codes in Computational Science and Engineering*, Discrete Mathematics and Its Applications (Boca Raton), Chapman & Hall/CRC, Boca Raton, FL, 2003. (Cited on p. 79)

[111] A. N. KOLMOGOROV AND S. V. FOMIN, *Elements of the Theory of Functions and Functional Analysis. Vol. 1. Metric and Normed Spaces*, Graylock Press, Rochester, N. Y., 1957. Translated from the first Russian edition by Leo F. Boron. (Cited on p. 63)

[112] D. KOURI, M. HEINKENSCHLOSS, D. RIDZAL, AND B. VAN BLOEMEN WAANDERS, *A trust-region algorithm with adaptive stochastic collocation for PDE optimization under uncertainty*, SIAM J. Sci. Comput., 35 (2013), pp. A1847–A1879. (Cited on p. 77)

[113] J. KRAUS AND S. MARGENOV, *Robust Algebraic Multilevel Methods and Algorithms*, vol. 5 of Radon Series on Computational and Applied Mathematics, Walter de Gruyter GmbH & Co. KG, Berlin, 2009. (Cited on p. 70)

[114] A. KUFNER, O. JOHN, AND S. FUČÍK, *Function Spaces*, Noordhoff International Publishing, Leyden, 1977. (Cited on p. 13)

[115] A. B. J. KUIJLAARS, *Convergence analysis of Krylov subspace iterations with methods from potential theory*, SIAM Rev., 48 (2006), pp. 3–40. (Cited on p. 75)

[116] A. KUNOTH, *Multilevel preconditioning—appending boundary conditions by Lagrange multipliers*, Adv. Comput. Math., 4 (1995), pp. 145–170. (Cited on p. 59)

[117] C. LANCZOS, *An iteration method for the solution of the eigenvalue problem of linear differential and integral operators*, J. Research Nat. Bur. Standards, 45 (1950), pp. 255–282. (Cited on p. 43)

[118] ———, *Solution of systems of linear equations by minimized iterations*, J. Research Nat. Bur. Standards, 49 (1952), pp. 33–53. (Cited on p. 43)

[119] ———, *The Inspired Guess in the History of Physics*, Studies, 53 (1964), pp. 398–412. (Cited on pp. vii, viii, 84)

[120] ———, *Linear differential operators*, Dover Publications, Inc., Mineola, NY, 1997. Reprint of the 1961 original. (Cited on p. viii)

[121] P. D. LAX AND A. N. MILGRAM, *Parabolic equations*, in Contributions to the Theory of Partial Differential Equations, Annals of Mathematics Studies, no. 33, Princeton University Press, Princeton, N. J., 1954, pp. 167–190. (Cited on p. 27)

[122] J. LERAY, *Sur le mouvement d'un liquide visqueux emplissant l'espace*, Acta Math., 63 (1934), pp. 193–248. (Cited on pp. 11, 14)

[123] J. LIESEN, *Pták's nondiscrete induction and its application to matrix iterations*, ArXiv:1405.2683, 2014. (Cited on p. 75)

[124] J. LIESEN AND Z. STRAKOŠ, *Krylov Subspace Methods: Principles and Analysis*, Numerical Mathematics and Scientific Computation, Oxford University Press, Oxford, 2013. (Cited on pp. 39, 42, 43, 44, 45, 46, 56, 59, 69, 70, 74, 75, 78, 79, 80, 82)

[125] J.-L. LIONS AND E. MAGENES, *Non-Homogeneous Boundary Value Problems and Applications. Vol. I*, Springer-Verlag, New York, 1972. Translated from the French by P. Kenneth. (Cited on p. 13)

[126] J.-L. LIONS AND G. STAMPACCHIA, *Variational inequalities*, Comm. Pure Appl. Math., 20 (1967), pp. 493–519. (Cited on p. 27)

[127] D. G. LUENBERGER, *Optimization by Vector Space Methods*, John Wiley & Sons Inc., New York, 1969. (Cited on pp. 38, 43)

[128] J. MÁLEK AND K. R. RAJAGOPAL, *Compressible generalized Newtonian fluids*, Z. Angew. Math. Phys., 61 (2010), pp. 1097–1110. (Cited on p. 19)

[129] J. MANDEL AND R. TEZAUR, *Convergence of a substructuring method with Lagrange multipliers*, Numer. Math., 73 (1996), pp. 473–487. (Cited on p. 77)

[130] K.-A. MARDAL AND R. WINTHER, *Preconditioning discretizations of systems of partial differential equations*, Numer. Linear Algebra Appl., 18 (2011), pp. 1–40. (Cited on pp. 30, 31, 32)

[131] I. MAREK AND K. ŽITNÝ, *Matrix Analysis for Applied Sciences. Vol. 1*, vol. 60 of Teubner-Texte zur Mathematik [Teubner Texts in Mathematics], BSB B. G. Teubner Verlagsgesellschaft, Leipzig, 1983. With German, French and Russian summaries. (Cited on p. 44)

[132] ——, *Matrix Analysis for Applied Sciences. Vol. 2*, vol. 84 of Teubner-Texte zur Mathematik [Teubner Texts in Mathematics], BSB B. G. Teubner Verlagsgesellschaft, Leipzig, 1986. With German, French and Russian summaries. (Cited on p. 44)

[133] W. MCLEAN AND T. TRAN, *A preconditioning strategy for boundary element Galerkin methods*, Numer. Methods Partial Differential Equations, 13 (1997), pp. 283–301. (Cited on p. 30)

[134] G. MEURANT, *Computer Solution of Large Linear Systems*, vol. 28 of Studies in Mathematics and its Applications, North-Holland Publishing Co., Amsterdam, 1999. (Cited on pp. 55, 58)

[135] ——, *The Lanczos and Conjugate Gradient Algorithms. From Theory to Finite Precision Computations*, vol. 19 of Software, Environments, and Tools, SIAM, Philadelphia, PA, 2006. (Cited on p. 75)

[136] ——, *GMRES and the Arioli, Pták, and Strakoš parametrization*, BIT, 52 (2012), pp. 687–702. (Cited on p. 75)

[137] G. MEURANT AND Z. STRAKOŠ, *The Lanczos and conjugate gradient algorithms in finite precision arithmetic*, Acta Numer., 15 (2006), pp. 471–542. (Cited on pp. 75, 78)

[138] G. MEURANT AND P. TICHÝ, *On computing quadrature-based bounds for the A-norm of the error in conjugate gradients*, Numerical Algorithms, 62 (2013), pp. 163–191. (Cited on p. 78)

[139] P. MORIN, R. H. NOCHETTO, AND K. G. SIEBERT, *Convergence of adaptive finite element methods*, SIAM Rev., 44 (2002), pp. 631–658. Revised reprint of "Data oscillation and convergence of adaptive FEM" [SIAM J. Numer. Anal. 38 (2000), pp. 466–488]. (Cited on p. 80)

[140] J. NAZARETH, *Conjugate gradient methods less dependent on conjugacy*, SIAM Review, 28 (1986), pp. 501–511. (Cited on p. 38)

[141] J. NEČAS, *Direct Methods in the Theory of Elliptic Equations*, Springer Monographs in Mathematics, Springer, Heidelberg, 2012. Translated from the 1967 French original. (Cited on pp. 7, 10, 16, 18)

[142] J. NEČAS AND I. HLAVÁČEK, *Mathematical Theory of Elastic and Elasto-Plastic Bodies: An Introduction*, vol. 3 of Studies in Applied Mechanics, Elsevier Scientific Publishing Co., Amsterdam-New York, 1980. (Cited on p. 3)

[143] O. NEVANLINNA, *Convergence of iterations for linear equations*, Lectures in Mathematics ETH Zürich, Birkhäuser Verlag, Basel, 1993. (Cited on p. 75)

[144] J. T. ODEN AND S. PRUDHOMME, *Goal-oriented error estimation and adaptivity for the finite element method*, Comput. Math. Appl., 41 (2001), pp. 735–756. (Cited on p. 78)

[145] C. W. OSEEN, *Neuere Methoden Und Ergebnisse in Der Hydrodynamik*, Akademische Verlagsgesellschaft M. B. H., Leipzig, 1927. (Cited on p. 14)

[146] P. OSWALD, *Stable subspace splittings for Sobolev spaces and domain decomposition algorithms*, in Domain Decomposition Methods in Scientific and Engineering Computing (University Park, PA, 1993), vol. 180 of Contemp. Math., Amer. Math. Soc., Providence, RI, 1994, pp. 87–98. (Cited on p. 59)

[147] J. PAPEŽ, J. LIESEN, AND Z. STRAKOŠ, *Distribution of the discretization and algebraic error in numerical solution of partial differential equations*, Linear Algebra Appl., 449 (2014), pp. 89–114. (Cited on pp. 39, 59, 69, 79, 82)

[148] A. T. PATERA AND E. M. RØNQUIST, *A general output bound result: application to discretization and iteration error estimation and control*, Math. Models Methods Appl. Sci., 11 (2001), pp. 685–712. (Cited on p. 78)

[149] J. W. PEARSON, *Fast iterative solvers for PDE-constrained optimization problems*, PhD thesis, University of Oxford, 2013. (Cited on p. 30)

[150] J. W. PEARSON AND A. J. WATHEN, *Fast iterative solvers for convection-diffusion control problems*, Electron. Trans. Numer. Anal., 40 (2013), pp. 294–310. (Cited on p. 30)

[151] D. PELLETIER AND P. J. ROACHE, *Verification and validation of computational heat transfer*, in Handbook of Numerical Heat Transfer, W. J. Minkowycz, E. M. Sparrow, and J. Y. Murthy, eds., John Wiley & Sons, Inc., New York, second ed., 2009, pp. 417–442. (Cited on p. 79)

[152] CH. PFLAUM, *Algebraic analysis of multigrid algorithms*, Numer. Linear Algebra Appl., 6 (1999), pp. 701–728. (Cited on p. 59)

[153] M. PICASSO, *A stopping criteria for the conjugate gradient algorithm in the framework of adaptive finite elements*, Comm. Numer. Meth. Eng., 25 (2009), pp. 339–355. (Cited on p. 78)

[154] C. E. POWELL AND D. J. SILVESTER, *Optimal preconditioning for Raviart–Thomas mixed formulation of second-order elliptic problems*, SIAM J. Matrix Anal. Appl., 25 (2003), pp. 718–738. (Cited on p. 30)

[155] V. PRŮŠA AND K. R. RAJAGOPAL, *Jump conditions in stress relaxation and creep experiments of Burgers type fluids: study in the application of Colombeau algebra of generalized functions*, Z. Angew. Math. Phys., 62 (2011), pp. 707–740. (Cited on p. 14)

[156] ———, *On implicit constitutive relations for materials with fading memory*, J. Non-Newtonian Fluid Mech., 181 (2012), pp. 22–29. (Cited on p. 19)

[157] V. PTÁK, *What should be a rate of convergence?*, RAIRO Anal. Numér., 11 (1977), pp. 279–286. (Cited on p. 75)

[158] I. PULTAROVÁ, *Preconditioning of the coarse problem in the method of balanced domain decomposition by constraints*, Math. Comput. Simulation, 82 (2012), pp. 1788–1798. (Cited on p. 59)

[159] A. QUARTERONI, R. SACCO, AND F. SALERI, *Numerical Mathematics*, vol. 37 of Texts in Applied Mathematics, Springer-Verlag, Berlin, second ed., 2007. (Cited on p. 65)

[160] K. R. RAJAGOPAL, *On implicit constitutive theories*, Appl. Math., 48 (2003), pp. 279–319. (Cited on p. 19)

[161] ———, *On implicit constitutive theories for fluids*, J. Fluid Mech., 550 (2006), pp. 243–249. (Cited on p. 19)

[162] ———, *The elasticity of elasticity*, Z. Angew. Math. Phys., 58 (2007), pp. 309–317. (Cited on p. 19)

[163] ———, *On the nonlinear elastic response of bodies in the small strain range*, Acta Mechanica, 225 (2014), pp. 1545–1553. (Cited on p. 19)

[164] K. R. RAJAGOPAL AND A. R. SRINIVASA, *On the response of non-dissipative solids*, Proc. R. Soc. A, 463 (2007), pp. 357–367. (Cited on p. 19)

[165] ———, *On the thermodynamics of fluids defined by implicit constitutive relations*, Z. Angew. Math. Phy., 59 (2008), pp. 715–729. (Cited on p. 19)

[166] ———, *On a class of non-dissipative materials that are not hyperelastic*, Proc. R. Soc. A, 465 (2009), pp. 493–500. (Cited on p. 19)

[167] R. RANNACHER, *Error control in finite element computations. An introduction to error estimation and mesh-size adaptation*, in Error Control and Adaptivity in Scientific Computing (Antalya, 1998), vol. 536 of NATO Sci. Ser. C Math. Phys. Sci., Kluwer Acad. Publ., Dordrecht, 1999, pp. 247–278. (Cited on pp. 78, 79)

[168] ———, *A short course on numerical simulation of viscous flow: discretization, optimization and stability analysis*, Discrete Contin. Dyn. Syst. Ser. S, 5 (2012), pp. 1147–1194. (Cited on pp. 78, 79)

[169] S. I. REPIN, *A Posteriori Estimates for Partial Differential Equations*, vol. 4 of Radon Series on Computational and Applied Mathematics, Walter de Gruyter GmbH & Co. KG, Berlin, 2008. (Cited on p. 78)

[170] P. J. ROACHE, *Quantification of uncertainty in computational fluid dynamics*, in Annual Review of Fluid Mechanics, vol. 29, Annual Reviews, Palo Alto, CA, 1997, pp. 123–160. (Cited on p. 79)

[171] ———, *Verification and Validation in Computational Science and Engineering*, Hermosa Publishers, Albuquerque, New Mexico, 1998. (Cited on p. 79)

[172] Y. SAAD, *Iterative Methods for Sparse Linear Systems*, SIAM, Philadelphia, PA, second ed., 2003. (Cited on pp. 55, 58)

[173] A. DE SAINT-EXUPÉRY, *The Wisdom of the Sands*, Amereon House, Mattituck, NY, 1952. Reprint of the publication by Hollis&Carter Ltd., London. (Cited on p. vii)

[174] V. K. SAUL'YEV, *Integration of Equations of Parabolic Type by the Method of Nets*, vol. 54 of International Series of Monographs in Pure and Applied Mathematics, Pergamon Press, London, 1960. Translated from the Russian by G. J. Tee, translation edited and editorial introduction by K. L. Stewart. (Cited on p. 15)

[175] D. J. SILVESTER AND A. J. WATHEN, *Fast iterative solution of stabilised Stokes systems. II. Using general block preconditioners*, SIAM J. Numer. Anal., 31 (1994), pp. 1352–1367. (Cited on p. 30)

[176] D. J. SILVESTER AND V. SIMONCINI, *An optimal iterative solver for symmetric indefinite systems stemming from mixed approximation*, ACM Trans. Math. Software, 37 (2011), pp. Art. 42, 22. (Cited on p. 78)

[177] O. STEINBACH AND W. L. WENDLAND, *The construction of some efficient preconditioners in the boundary element method*, Adv. Comput. Math., 9 (1998), pp. 191–216. (Cited on p. 30)

[178] R. STEVENSON, *Optimality of a standard adaptive finite element method*, Found. Comput. Math., 7 (2007), pp. 245–269. (Cited on p. 80)

[179] T. J. STIELTJES, *Recherches sur les fractions continues*, Ann. Fac. Sci. Toulouse Sci. Math. Sci. Phys., 8 (1894), pp. J. 1–122. Reprinted in Oeuvres II (P. Noordhoff, Groningen, 1918), pp. 402–566. English translation *Investigations on continued fractions* in Thomas Jan Stieltjes, Collected Papers, Vol. II (Springer-Verlag, Berlin, 1993), pp. 609–745. (Cited on pp. 43, 44)

[180] M. STOLL AND A. J. WATHEN, *Preconditioning for partial differential equation constrained optimization with control constraints*, Numer. Linear Algebra Appl., 19 (2012), pp. 53–71. (Cited on p. 30)

[181] ———, *All-at-once solution of time-dependent Stokes control*, J. Comput. Phys., 232 (2013), pp. 498–515. (Cited on p. 30)

[182] Z. STRAKOŠ, *On the real convergence rate of the conjugate gradient method*, Linear Algebra Appl., 154/156 (1991), pp. 535–549. (Cited on pp. 74, 82)

[183] Z. STRAKOŠ AND J. LIESEN, *On numerical stability in large scale linear algebraic computations*, ZAMM Z. Angew. Math. Mech., 85 (2005), pp. 307–325. (Cited on p. 75)

[184] Z. STRAKOŠ AND P. TICHÝ, *On error estimation in the conjugate gradient method and why it works in finite precision computations*, Electron. Trans. Numer. Anal., 13 (2002), pp. 56–80. (Cited on pp. 75, 78)

[185] R. TEMAM, *Problèmes Mathématiques en Plasticité*, vol. 12 of Méthodes Mathématiques de l'Informatique, Gauthier-Villars, Montrouge, 1983. (Cited on p. 20)

[186] P. S. VASSILEVSKI, *Multilevel Block Factorization Preconditioners: Matrix-Based Analysis and Algorithms for Solving Finite Element Equations*, Springer, New York, 2008. (Cited on p. 59)

[187] R. VERFÜRTH, *A Posteriori Error Estimation Techniques for Finite Element Methods*, Numerical Mathematics and Scientific Computation, Oxford University Press, Oxford, 2013. (Cited on p. 78)

[188] M. VOHRALÍK, *A Posteriori Error Estimates, Stopping Criteria, and Inexpensive Implementations for Error Control and Efficiency in Numerical Simulations*, Habilitation Thesis, Université Pierre et Marie Curie, Paris, 2010. (Cited on p. 78)

[189] J. VON NEUMANN, *Mathematische Begründung der Quantenmechanik*, Nachr. Ges. Wiss. Göttingen, Math.-phys. Kl., (1927), pp. 1–57. (Cited on p. 44)

[190] ———, *Mathematical Foundations of Quantum Mechanics*, Princeton Landmarks in Mathematics, Princeton University Press, Princeton, NJ, 1996. Translated from the 1932 German original and with a preface by R. T. Beyer. (Cited on p. 44)

[191] J. VON NEUMANN AND H. H. GOLDSTINE, *Numerical inverting of matrices of high order*, Bull. Amer. Math. Soc., 53 (1947), pp. 1021–1099. (Cited on pp. viii, 75)

[192] YU. V. VOROBYEV, *Methods of Moments in Applied Mathematics*, Translated from the Russian by Bernard Seckler, Gordon and Breach Science Publishers, New York, 1965. (Cited on pp. 42, 43, 44, 46, 47, 73)

[193] K. WASHIZU, *Variational Methods in Elasticity and Plasticity*, Pergamon Press, Oxford, second ed., 1975. With a foreword by R. L. Bisplinghoff, International Series of Monographs in Aeronautics and Astronautics, Division I: Solid and Structural Mechanics, Vol. 9. (Cited on p. 3)

[194] J. WLOKA, *Partial Differential Equations*, Cambridge University Press, Cambridge, 1987. Translated from the German by C. B. Thomas and M. J. Thomas. (Cited on pp. 7, 12)

[195] B. I. WOHLMUTH AND R. H. W. HOPPE, *A comparison of a posteriori error estimators for mixed finite element discretizations by Raviart-Thomas elements*, Math. Comp., 68 (1999), pp. 1347–1378. (Cited on p. 78)

[196] H. WU AND Z. CHEN, *Uniform convergence of multigrid V-cycle on adaptively refined finite element meshes for second order elliptic problems*, Sci. China Ser. A, 49 (2006), pp. 1405–1429. (Cited on p. 80)

[197] J. XU, *Iterative methods by space decomposition and subspace correction*, SIAM Rev., 34 (1992), pp. 581–613. (Cited on p. 59)

[198] J. XU AND L. ZIKATANOV, *Some observations on Babuška and Brezzi theories*, Numer. Math., 94 (2003), pp. 195–202. (Cited on p. 65)

[199] H. YSERENTANT, *Hierarchical bases of finite-element spaces in the discretizaton of nonsymmetric elliptic boundary value problems*, Computing, 35 (1985), pp. 39–49. (Cited on pp. 51, 59)

[200] ———, *Erratum: "On the multilevel splitting of finite element spaces,"* Numer. Math., 50 (1986), p. 123. (Cited on pp. 51, 59)

[201] ———, *Hierarchical bases give conjugate gradient type methods a multigrid speed of convergence*, Appl. Math. Comput., 19 (1986), pp. 347–358. Second Copper Mountain conference on multigrid methods (Copper Mountain, Colo., 1985). (Cited on pp. 51, 59)

[202] ———, *On the multilevel splitting of finite element spaces*, Numer. Math., 49 (1986), pp. 379–412. (Cited on pp. 51, 59)

[203] O. C. ZIENKIEWICZ, B. M. IRONS, F. E. SCOTT, AND J. S. CAMPBELL, *High speed computing of elastic structures*, in Proceedings of the Symposium of the International Union of Theoretical and Applied Mechanics, Liege, 1970. (Cited on p. 59)

[204] W. ZULEHNER, *Nonstandard norms and robust estimates for saddle point problems*, SIAM J. Matrix Anal. Appl., 32 (2011), pp. 536–560. (Cited on pp. 30, 31)

Index